GLOBAL
WARMING
AND POLITICAL
INTIMIDATION

GLOBAL
WARMING
AND POLITICAL
INTIMIDATION

How Politicians
Cracked Down on Scientists
as the Earth Heated Up

RAYMOND S. BRADLEY

UNIVERSITY OF MASSACHUSETTS PRESS
Amherst & Boston

LC 2011010043
ISBN 978-1-55849-869-3 (paper)
ISBN 978-1-55849-868-6 (library cloth)

Designed by Sally Nichols
Set in Minion Pro and Frutiger
Printed and bound by Thomson-Shore, Inc.

Library of Congress Cataloging-in-Publication Data

Bradley, Raymond S., 1948–
Global warming and political intimidation : how politicians cracked
down on scientists as the earth heated up / Raymond S. Bradley.
p. cm.
Includes bibliographical references and index.
ISBN 978-1-55849-869-3 (pbk. : alk. paper) — ISBN 978-1-55849-868-6
(library cloth : alk. paper) 1. Global warming—Political aspects.
2. Climate change mitigation—Political aspects. 3. Science—Political
aspects. 4. Politics and culture. I. Title.
QC981.8.G56B7 2011
577.27´6—dc22
2011010043

British Library Cataloguing in Publication data are available.

....

To Congressman Sherwood Boehlert,
who put principle before politics,
and set an example for others to follow.

....

CONTENTS

ILLUSTRATIONS

ACKNOWLEDGMENTS

My thanks to the editors Brian Halley and Amanda Heller at University of Massachusetts Press for their wise counsel in the editorial process, and to family members, friends, and anonymous reviewers, whose sage advice I tried to adopt in revising my original text. Thanks also to Kinuyo Kanamaru for assistance with the figures.

GLOBAL
WARMING
AND POLITICAL
INTIMIDATION

PROLOGUE

It is often said that global warming is the number one environmental issue of our time. Some experts go even further, arguing that it threatens civilization as we know it. As a working climatologist for over thirty-five years, I have seen an explosive growth in our understanding of global climate and the increasing role that human activity plays in controlling the earth's climate system. By our extravagant use of fossil fuels (coal, oil, and gas), we are rapidly returning to the atmosphere carbon dioxide (CO_2) that was extracted by plants millions of years ago. This has already affected the energy balance of the earth, but the changes we can expect later in this century will be far greater unless fossil fuel emissions are curbed.

Prior to the Industrial Revolution, carbon dioxide levels in the atmosphere were about 275 parts per million by volume (ppmv); by the time I started graduate school in 1969, they had risen to 320 ppmv, and by 2010 they had reached about 390 ppmv. Most of this increase has been due to the burning of fossil fuels by the wealthy, more developed countries of the world. But the rate at which carbon dioxide is accumulating in the atmosphere is rising ever more rapidly. Part of this increase has to do with world population growth, which has accelerated dramatically over the past fifty years or so, but a great deal of it is related to strong economic development in the countries of what used to be known as the Third World. "Third World" no longer, these are now the fastest-growing economies on the planet.

When I first went to China in 1984, the entire city of Beijing was dark at night. There was not a light anywhere. Today the city can be seen aglow at night even from a satellite. In 1984 virtually everyone got around by bicycle. Electricity was rationed; I arrived at a university on a Thursday to give a lecture (with slides), only to learn that it didn't get electricity on Thursdays. As I write, barely twenty-five years later, China has been completely transformed, with millions of cars on an ever-expanding highway system, high-speed trains connecting its cities, and thousands of high-rise apartment buildings offering condominiums equipped with all the latest appliances and technological gadgets. A new coal-burning power plant opens somewhere in China every month. And pretty much the same thing is happening across South Asia, Russia, and South America.

All of these changes have driven fossil fuel consumption to record heights, with dramatic changes in greenhouse gas levels throughout the world. You can stand at the South Pole and measure the effect of this economic growth taking place tens of thousands of miles away. The problem is truly one of global proportions, and the solution will require a global effort.

None of this was very obvious as I completed my graduate studies in climatology in the early 1970s. I worked in the Arctic, more focused on glacial advances in the past than the changes in climate that were starting to occur. But technological changes were taking place in my own science, too. Dramatic advances in climate modeling made it possible to simulate with remarkable accuracy the climate of the present day as well as climates in the past, which we could verify with paleoclimate records. These same models provide a worrisome picture of how future climate will evolve if the concentration of carbon dioxide and other greenhouse gases in the atmosphere continues to rise.

Just as the field of climatology has evolved since the 1970s, so too has my own thinking about the problem of global warming. I was never an evangelist about the issue. But as I've learned more about the subject, and scientific evidence has accumulated, like almost every other climatologist on the planet, I've become convinced that global warming is a critical issue that requires urgent attention. Reinforcing this view are the periodic assessments of the "state of the science" by the Intergovernmental Panels on Climate Change (IPCC), which are made up of numerous scientists (including me) brought together to review the published literature and produce summaries of where things stand. Each IPCC report represents the best effort of scientists working together to figure out such issues as carbon emissions, ongoing climate change, and the actual and potential impacts of future changes under different potential scenarios. It's hard to imagine a process that could be more open or fair, given that just about anybody can read the initial draft reports and provide critical comments, all of which must be evaluated and responded to. Nevertheless, the IPCC and the science surrounding global warming have become embroiled in politics, and nowhere more so than in the United States.

This book describes my own involvement in this issue and my personal experiences at the center of attempts by conservative politicians to dismiss the significance of human-induced (anthropogenic) global warming. Their strategy was to attack the results of my research on climate change as well as the research of my colleagues. But these attempts went far beyond simple disagreements about the science; these policymakers sought to discredit my reputation and that of other climatologists, and to intimidate us, and by extension anybody else who might be considering publishing results that were unwelcome on Capitol Hill. I became caught up in this sleazy campaign because of the so-called "hockey stick" graph, which my colleagues Michael Mann and Malcolm Hughes and I published in the journal *Nature* in 1998, with follow-on studies the next year. Our study concluded that temperatures in recent decades were the warmest in a thousand years. This might not sound earth-shattering now, but in the late 1990s it so incensed the entrenched interests in Congress that they used their political power to try to tarnish our scientific credentials and bury us under so many demands for information that, if we had complied, our research would essentially have come to a halt. Angry about the message of our research, they sought to destroy the reputations of the messengers. We were not the first victims of the attack on climate science. That dubious honor falls to my friend Ben Santer, who had been vilified a few years earlier for his work on the 1995 IPCC report (discussed in chapter 4). Nor were we the last, as I discuss in the final chapter. Numerous scientists have been swept up in the dirty politics of global warming.

In retrospect, it seems remarkable that our hockey stick study would have generated a political backlash, but such were the stakes. Concern was mounting that the public might actually wake up and demand legislative action that would limit energy consumption and associated profits for the energy industry. The hockey stick graph was chosen as the sacrificial lamb. Our antagonists seemed to believe that if they could refute our study, the entire edifice of global warming science would crumble and fall. Nothing could have been further from the truth, as concern over global warming rests on a vast array of scientific evidence, of which the hockey stick is but a minuscule part. The fact that our critics seized on this particular study to launch their attack simply underlines their ignorance about the essential facts. If we had never published the graph, the issue of global warming would be no less urgent or compelling. But this incident became much larger than an assault on our integrity; it engaged politics and science in a head-to-head battle that threatened the very nature of the way scientific research is conducted (and supported) in the United States. This was quickly recognized by numerous scientific organizations, which weighed in not to defend our results but to defend the way in which science proceeds—driven not by political expediency but by critical scientific evaluation over many years.

This book, then, is an account of what we know about climate change, and my experiences as a climate scientist who became unwittingly caught up in a controversy that was inflated to a far greater magnitude than the underlying science justified, simply because vested interests perceived a threat and sought to eliminate it. Our opponents responded by using the power of state institutions to intimidate us as individual scientists. Such a strategy was part of a larger pattern involving government suppression of scientific information to ensure that the "party line" was promoted at the expense of truthful scientific discourse. Such tactics are not worthy of an open and democratic society. The hockey stick scientific debate is over, but government interference in science at the local, state, and federal level remains a contentious issue in the United States and must not be ignored.

To paraphrase Andrew Jackson's Farewell Address of 1837, "Eternal vigilance by the people is the price of [scientific] liberty, and . . . you must pay the price if you wish to secure the blessing. It behooves you, therefore, to be watchful in your States as well as in the Federal Government."

1
The Congressional Hearings
The Good, the Bad, and the Ugly

The Russell Office Building on Capitol Hill is an imposing structure with long, high-ceilinged corridors that might remind you of a cathedral. Clusters of people scurry by, all in a hurry, all on important business no doubt. This is where many senators have their offices, each suite plush and patriotic, with flags and photographs of the senator shaking hands with the president or some other head of state, or meeting the troops. Walls are plastered with letters of thanks, honorary degrees from universities (thankful for past earmarks), letters of endorsement from influential organizations, and more photographs of people I suppose I should know but usually don't. This is as close as the average citizen gets to the power and glory of Washington's political elite.

It's easy to see how seductive the political process can be; in here, politicians rule. But as each new election approaches, money—lots of money—has to be raised to ensure reelection, so the implicit obligation to those who provide reliable support grows stronger. The longer a senator stays in Congress, the greater the influence accrued, increasing his or her ability to deliver favorable legislation, subsidies, or tax breaks. This growing power in turn attracts further support. It's no surprise that historically over 90 percent of incumbents have been reelected. In climate science we refer to this as positive feedback; once you start a process going, it tends to gain a life of its own as additional factors, some perhaps unforeseen, begin to operate to reinforce it.

I expect that many of these politicians never knew how many friends they had until they were elected and appointed to the Senate committee overseeing legislation—on farm subsidies, or transportation infrastructure, or energy. If you're a friend of the energy industry and in an

7

influential position in Congress, is there any way you could lose an election? Just how valuable would it be to you to be Big Energy's number one friend in the U.S. Congress?

These thoughts swam through my mind on May 17, 2000, as I wandered down the long corridors of the Russell Building looking for the room where I had been invited to testify at a hearing on "the science behind global warming," before the Senate Committee on Commerce, Science, and Transportation. The chairman of the committee was Republican senator John McCain of Arizona, and the ranking minority member was Democratic senator John Kerry of Massachusetts. Normally the chairman selects the principal witnesses at a hearing. As a result, it is the views and perspective of the party in charge that get entered into the *Congressional Record,* and whatever publicity the hearing generates will generally spin in the preferred direction. In this case I was told that Senator McCain had invited Senator Kerry to recommend witnesses. This seemed a promising start, as it suggested that the hearing might actually be a genuine attempt by the members of the committee to obtain useful information rather than just a political stunt. It didn't take very long for this view to be confirmed. Senator McCain opened the proceedings by noting that in his 1999 run for president, everywhere he went he saw people holding signs asking about his plans to combat global warming. "In town hall meeting after town hall meeting after town hall meeting, of which I had hundreds," he recalled, "young Americans stood up and said, 'Senator McCain, what is your position on global warming? . . . What is your plan?'"[1] The campaign to which he was referring was organized by an environmental group known as Ozone Action, which tried to make sure that wherever candidates for president were campaigning, there would be somebody holding a sign or asking questions about global warming. This strategy had clearly worked.

McCain continued:

I am sorry to say that I do not have a plan because I do not have, nor do the American people have, sufficient information and knowledge. But I do believe that Americans and we who are policymakers in all branches of government, should be concerned about mounting evidence that indicates that something is happening. I do not pretend to have the expertise and knowledge on this very important and very controversial issue, but I do intend, beginning with this hearing and follow-on hearings, to become informed, to reach some conclusions, and make some recommendations. . . . I believe that it is of the utmost importance that we examine this issue thoroughly, and I am dedicated to that proposition.

To me, this was an amazing moment. Here was one of the country's leading politicians, openly admitting his lack of knowledge about global warming and asking for help. I was very impressed by what seemed an open and honest start to the proceedings and was very glad that I had accepted Senator Kerry's invitation.

In fact, I was one of five scientists who had been asked to testify. The others were Neal Lane, assistant to the president for science and technology in the Clinton White House's Office of Science and Technology Policy; Jerry Mahlman, director of the Geophysical Fluid Dynamics Lab of the National Oceanic and Atmospheric Administration (NOAA); Kevin Trenberth, director of the Climate Analysis Section of the National Center for Atmospheric Research; Robert Watson, chairman of the IPCC; and John Christy, director of the Earth Systems Science Center at the University of Alabama. No fluff there. This was a stellar cast, well qualified to cover all aspects of the climate change issue.

The proceedings continued with each senator making a statement. Senator Kerry reflected the views of most of us waiting to testify:

I must say, Mr. Chairman, I have been a little bit dumbfounded and somewhat disturbed by the level of skepticism that exists, and has existed over a long period of time, in the U.S. Congress with respect to this issue. . . . [W]e know the worldwide rise in temperatures at the earth's surface is real. We know it has accelerated in recent decades. . . . Anybody who does not see the impact of these changes is putting their head in the sand. Now, can we say that every bit of this is due to global warming? The answer is no. I cannot sit here and tell you that. No scientist is going to tell you that every bit of it is. Some of it may be normal changes that are taking place in terms of the climate process, but we do know with absolute certainty, incontrovertible scientific fact, we are contributing to it. And we ought to adopt the prudent person theory with respect to those things that you do not quite know what the final consequences are going to be, but you know they might be disastrous.

I thought this was an excellent point. Why wouldn't you take modest steps to reduce your risk of something bad happening? We mandate that everyone driving a car must have insurance in case of an accident, because we recognize that risk. Most homeowners have a fire insurance policy, yet few ever expect to suffer the consequences of a catastrophic fire. Shouldn't we seek legislation that provides the country and the world a measure of insurance against catastrophic risk, even if that possibility is uncertain? How much has the United States spent on a "Star Wars" antiballistic missile shield to protect against an uncertain risk with a very low probability?

Kerry then brought things down to a more prosaic level. "I remember as a kid in Massachusetts," he said, "we always looked forward to October/November because the ponds froze over and we were going to have thick ice, and go play hockey. Today, you are lucky if the ponds freeze in Northern New Hampshire." Ah, yes, all politics is local. Nobody cares too much about *global* climate change, but once the effects hit home locally, they have a much greater impact. A few minutes later, Senator Sam Brownback, the Kansas Republican, brought this point home once again by providing his own perspective: "As Senator Kerry was talking about how he looked forward to the winter and playing ice hockey, I was sitting here thinking I was out cutting holes in the ice to water cattle, and I did not like that, as thick as it got. It is not that I am saying we should have global warming. I do not agree with that. I am not for global warming, but we just did not play the ice hockey. I had to cut ice holes."

Here was the issue brought down to its essential point: not everybody would necessarily be unhappy if it was a bit warmer (in winter at least). I could appreciate those sentiments. I remember walking my dog around the block one bitterly cold January day in western Massachusetts, where I live, struggling against an icy wind and wondering to myself how anybody could be persuaded that global warming was such a bad thing. The problem is that you can't accept just one part of the global warming story (milder winters) without getting some of the other attributes (hotter summers, more extreme weather, rising sea levels, and so on). Perhaps I was being naïve, but I got the impression that Brownback and the other senators could concede that argument. They seemed open-minded and concerned about the problem. Kerry concluded by noting that, far from endorsing the Kyoto Protocol, an international agreement to limit greenhouse gas emissions, Congress had actually forbidden any participation in further discussion of it. "[L]anguage . . . in the Agricultural Appropriations Bill," he warned, "will limit the Administration's activities on an international level to even continue the dialogue and process of building a consensus about Kyoto." Kerry's point here showed how far to the right the Congress had moved on this issue by the end of the Clinton administration. It wasn't enough to ignore the issue of global warming. Congress took the position that the United States should not even discuss it! Like a spoiled kid, it had grabbed all the toys and refused to play. Neal Lane, President Clinton's science adviser, summed up his concerns:

> The rider [to the Agricultural Appropriations Bill] seems, on the face of it, extreme. It tries to block the United States from even trying to reach an agreement with other countries on action to combat global warming, which is very difficult to explain to our international partners around the world. It undermines the ability of the executive branch to conduct international negotiations, which seems to me to raise serious constitutional questions. It may stifle U.S. efforts to achieve bi-partisan goals with a cost effective treaty and meaningful participation of developing countries.

Congress's take-no-prisoners approach to legislation was a reflection of how extreme views had managed to take center stage in U.S. politics, led by industry lobbyists who had captured the congressional legislative process. Not long after this hearing, in 2003, legislation was passed that expressly prohibited government agencies involved in Medicare prescription drug purchases from negotiating with pharmaceutical firms for lower rates. Who stood to gain in that lobbying coup? And who would gain from expressly forbidding the U.S. government to discuss limitations on carbon emissions?

When it came time for me to speak, I tried to grab the senators' attention right from the start. I began:

> *We are living in unusual times.* The climate of the twentieth century was dominated by universal warming; almost all parts of the earth had temperatures at the end of the century that were higher than when it began. At the same time, the concentration of greenhouse gases in the atmosphere increased to levels that were higher than at any time in *at least* the last 420,000 years. *These observations are incontrovertible.* Global warming is real and the levels of greenhouse gases (such as carbon dioxide) are now 35–40% higher than they were in the middle of the 19th century. This change in greenhouse gas concentration is largely the result of fossil fuel combustion.

My stark assessment seemed to have the desired effect. I looked up, and everyone was listening. I then tried to give a fair assessment of the state of the science, with some background on how climate had changed leading up to the present day (a topic I deal with in more depth in chapters 3 and 5). One of the great things about studying past climate change is that it enables you to put things in perspective. There wouldn't be much to worry about if the changes we're seeing today, or those we expect to see in the near future, were roughly equivalent to what we've experienced in the recent past. But unfortunately this is not the case, as I explained to the senators:

> The latest IPCC model-based projections of future climate point to a temperature increase of 0.6 to 2.2°C (~1 to 4°F) above 1990 levels by 2050. Clearly, these estimates have large uncertainties, but it is important to note that even the lowest value would be far beyond the range of temperatures in the last millennium. If these

estimates are even close to being correct, we are heading into uncharted waters relative to the climate of the last 1000 years.

I continued, "Why should we be concerned about global contamination of the atmosphere and future changes in climate?":

> [T]he levels of two important greenhouse gases (carbon dioxide and methane) [are] now higher than at any time in the last 420,000 years . . . based on measurements from the longest ice core record available (from the Russian Vostok station in Antarctica). . . . To put this in perspective, recall that it was only 10,000 years ago that human society first developed agriculture, and 120,000 years ago sabre-toothed tigers roamed what is now Trafalgar Square. Yet carbon dioxide levels have risen from fairly steady background levels (~270 ppmv) to present day levels (370 ppmv) in a little over a century. This rate of change has no parallel in the historical past, just as temperatures recorded in the late 20th century were unprecedented.

One of the problems we face in talking about global warming is to get across the issue of time and rates of change. If you say that something happened tens of thousands of years ago, it's hard for people to relate to that. You need to peg the discussion onto a timescale they can deal with. I'd hoped that the image of saber-toothed tigers living in what later became central London would help frame the immense amount of time these ice core records represent. In fact, since this hearing, ice core records of carbon dioxide have been extended even further back in time, over the last 850,000 years, and in all of that time we see no evidence that the atmosphere has ever contained carbon dioxide levels as high as they are today. This reality raises the question: Where did the carbon dioxide come from? I addressed this point next: "Most of the change in CO_2 and other greenhouse gases resulted from the growth of world population and the insatiable demand for fossil fuel–based energy. Given that world population will almost certainly double within the lifetime of those currently in kindergarten, unless something is done to curb the use of fossil fuel consumption, it seems very likely that significant changes in climate will occur in the near future." I looked across the room at the assembled panel of senators—all old enough to be grandparents. Maybe this reference to their children's children would register. But how to galvanize them into action? I felt as if I were in a small rubber dinghy trying to turn an enormous battleship in a different direction.

Can we be certain that future climate will involve unprecedented risks? Some argue that processes within the climate system will act to compensate for the effects of higher greenhouse gas levels (so-called negative feedback effects). According to this scenario, these feedbacks will help maintain the climatic *status quo*[,] enabling us to continue to contaminate the atmosphere *ad infinitum*. There is a small chance that such critics are right, in which case it would be safe to do nothing. But they may be completely wrong, and indeed the scientific consensus is that they *are* wrong.

Then it was time for a bit of backslapping encouragement: Come on, guys, let's go! Please don't just sit there. Do something!

Political decisions inevitably involve assessing risk and weighing the consequences of action versus inaction. Just as Congress must decide if the (perhaps small) risk of a rogue nation launching a nuclear missile at the United States (resulting in a catastrophe) is worth avoiding by spending large sums of money on a space defense system, so it must weigh the potentially catastrophic environmental and commercial consequences of future global warming against the costs of curbing fossil fuel consumption to reduce these risks. Scientists cannot provide Congress with a *certain* forecast of the future and as research on global warming continues, our understanding will undoubtedly change. But the picture at present is that we are indeed living in climatically unusual times, and that the future is likely to be even more unusual.

Amen. My testimony was over. But there was no crescendo of concern, no gavel pounding on the table with the chairman rising to restore order in the gallery as the crowd went wild. In fact, all in all it was a rather anticlimactic moment. Just polite thanks, move over, next speaker, please.

The next speaker turned out to be John Christy, professor of atmospheric science and director of the Earth System Science Center at the University of Alabama in Huntsville. John is one of a handful of scientists who have become known as "climate skeptics" or "contrarians," who are often invited to hearings such as this one to provide some sort of "balance." I wonder if, when the Space Science Committee meets to discuss the latest shuttle mission, they also invite somebody from the Flat Earth Society to ensure they are getting both sides of the story. At some point you have to accept that science has made some progress and move on. Gravity exists. The earth has a magnetic field. Evolution happened.

True to form, John reassured the senators that carbon dioxide is not a pollutant; it's just "plant food." And yes, the climate is changing, but "it always has and it always will." This is like saying to somebody who was just diagnosed with severe heart disease: "People have heart attacks all the time. Let's go for a quick jog around the Capitol Building." Not such a great idea, and not such great testimony either, in my view. Still, he performed as expected and no doubt supplied some cover for those senators present who needed something in the hearing that they could hide behind. But there was nothing in what the rest of us said that would provide much comfort for those hoping to dismiss it as scientific trivia. The warnings were clear, consistent, and stark: our climate is changing quickly, is likely to change soon to a new status that we have no experience of, and this change is largely driven by ever-increasing levels of greenhouse gases in the atmosphere. That's the bad news. The good news was that the senators present—most notably John McCain—listened, understood, and were persuaded.

Soon after, McCain became a global warming evangelist within the Republican Party (but only until his next re-election campaign), urging his recalcitrant colleagues to take the issue seriously, and—better yet—to pass legislation that would control carbon dioxide emissions. Yet the battleship analogy I alluded to earlier isn't too far off base. The ship of state is huge and unwieldy, and very slow to change direction. In this hearing we gave it an initial shove, and though it has yet to set an entirely new course, it is slowly turning in the right direction. Given the power of that enormous vessel, once it's on track, I feel confident that rapid progress toward a good solution can be achieved.

Although this hearing provided a certain level of satisfaction—at least some senators were paying attention—there were others in the same building who had an entirely different agenda. Led by Republican senator James Inhofe of Oklahoma, the Senate Committee on Environment and Public Works was laboring hard to dismiss the entire topic of global warming as pure fiction. In fact, in perhaps his most celebrated speech, Inhofe railed at scientists who simply reported the facts they observed, claiming that global warming was "the greatest hoax ever perpetrated on the American people" and that "man-induced global warming is an article of religious faith. . . . [T]he climate change debate should be based on fundamental principles of science, not religion. Ultimately, I hope, it will be decided by hard facts and data, and by serious scientists committed to the principles of sound science."[2]

The term Inhofe used, "sound science," has become a code word for those who reject the evidence linking greenhouse gases to global warming, or indeed for those who deny entirely that global warming is even occurring. Any science that supports their contrarian view is by definition "sound science"; all the rest is presumably a "hoax."

What makes Inhofe's rant particularly ironic is his attempt to dismiss the vast array of scientific evidence that has been published on anthropogenic climate change as nothing more than religious belief. In fact, when you start with the conviction that you know the truth and all who disagree are wrong, as Inhofe does, *that* is religion, not science. Good (and, yes, *really sound*) science does not mischaracterize past research to prove an article of faith.

Inhofe is very good at mischaracterizing science. He has been especially petulant about the "hockey stick." Over the years, he has been obsessed by it, ranting and raving about what a lie it is in virtually every speech he has made on global warming. In an appearance on Glenn

Beck's TV show (on July 20, 2006) Inhofe spluttered the following gibberish:

> The more I checked into it, the things started with the United Nations, the International Panel on Climate Control, and they used one scientist. And his name was Michael Mann, the famous hockey stick—remember that?—where he plotted the temperatures that went all the way across on a horizontal line, then you got to the twentieth century and it started going up. Well, one thing they forgot to do is put in the medieval warming period, which was from about 900 to 1400 AD, when it was warmer then than it is now. So . . . I was right on this thing. This thing is a hoax.[3]

It is always a good idea when trying to rally the fanatics to start with the United Nations (World government! Oppression! Freedom denied!). But "the International Panel on Climate Control"? That really is bad: the UN is trying to do away with the free will of the climate system! And as Inhofe revealed, there is only one scientist behind it, Mike Mann, the Darth Vader of the hockey team. But it gets worse. They ignored the "medieval warming period"! And everyone knows there was a "medieval warming period."

Here, of course, Inhofe stepped right into a trap of his own making. He claimed a priori that there was a medieval warm period, and so any and all research that does not recognize it must be wrong. In fact there *is* strong evidence that parts of Europe and the North Atlantic were warm for some portion of the Middle Ages, as my colleagues and I have discussed extensively in several published papers. But elsewhere in the world—for example, in the eastern equatorial Pacific, an area at least as large as western Europe—it was cold during similar intervals at that time. Not surprisingly, large-scale studies averaging out such records show that even if some areas were warm, other areas were cold, and so the net change across the globe was minimal.

What Inhofe fails to see is that his uncritical acceptance of popular notions refuting the global warming that he rails against merely highlights his dogmatic acceptance of what he considers "known truths."

Here is the warped logic he's following: We *know* there was a medieval warm period, so science that does not show it is obviously wrong, while the pseudoscience that denies global warming is obviously right, in spite of the overwhelming evidence to the contrary. Where does such confidence come from? What would stir the senator from Oklahoma, center of the U.S. oil and gas industry, to become such a vocal opponent of modern climate science? Could it be the $131,000 he received from labor groups between 2000 and 2008? Or would the $3 million he received from energy and natural resource industries have strengthened his resolve a little more firmly?[4]

Perhaps the most bizarre episode of Inhofe's sorry leadership of his Senate committee was the time when he held a hearing on "the role of science in environmental policy-making" (September 28, 2005).[5] This eloquent title was actually just a cover for more drumbeating by global warming's loudest disbeliever. But even by Inhofe's absurd standards, this particular hearing was especially odd, because the principal testimony came from a science-fiction writer, Michael Crichton. Inhofe lauded his guest, pouring praise on him like cream onto a peach cobbler. Somehow, Crichton's background in medicine had elevated his stature to that of Super Scientist: Have Problem, Will Testify. This notion that having some scientific training gives you carte blanche to weigh in on any issue is a relatively new phenomenon, possibly encouraged by the ubiquity of the Internet and facile access to blogging software. I really don't follow the logic. If I had a medical problem, I wouldn't want to be treated by a climatologist. So what possesses a doctor (an M.D., that is) to feel qualified to sound off about climate science is beyond me. As a fully paid-up climatologist of many years' standing, I know there is an immense amount about climate science that I *don't* know. The idea of weighing in on an entirely different field strikes me as presumptuous at best and foolish at worst.

But Dr. Crichton had no such concerns. He proceeded to hold forth on a wide range of topics, though most of his fire was reserved for the now infamous hockey stick graph that so enraged Inhofe and his followers. The work didn't show the "well-known" medieval warm period (here we go again) or even the Little Ice Age. Now, just for the record, the warmest hundred-year period in the hockey stick reconstruction (prior to the twentieth century) was 1084–1183, and the coldest was 1800–1899. Perhaps Crichton's medical training did not include interpreting graphs. No matter. Our research was phony, it was rubbish, it was apparently dismissed by "climate scientists around the world." Those who attempted to reproduce the research had been "obstructed at every turn" and told by the National Science Foundation (NSF) that the authors (Mike Mann, Malcolm Hughes, and I) "were under no obligation to provide . . . data to other researchers for independent verification."

Of course, none of this was true, though I suppose if you write science fiction for a living, fantasizing about reality comes naturally. In fact, the NSF never issued the statement ascribed to it, and we did indeed make data available; it was only Mike Mann's computer program—his private intellectual property—that was not provided. Other researchers, carefully following our procedures and writing their own computer code, quite independently produced results that were almost exactly the same as ours. And that, as Dr. Crichton solemnly noted, is the gold standard in science.

All this was grand theater, and as in many forms of light entertainment, the opening scene was the best part. Several senators fawned and fussed over the famous writer, basking in the warm glow that the distinguished celebrity brought to their normally mundane hearing. But others rose to the occasion. Democratic senator Barbara Boxer was deliciously sarcastic. "We are here to talk about sound science," she declared, nicely appropriating Inhofe's terminology. "We are not here to talk about plays, novels, art or music, although as a member for California, I do appreciate the focus on the arts." Her irony was probably far too subtle for those who were eager to hear words of wisdom from the great author.

Another Democratic senator, Frank Lautenberg of New Jersey, was more direct:

It might make a good story to imagine that the threat of global warming is a concoction of groups with a political agenda. But we need scientific facts . . . not science fiction. . . . By refusing to act, we are gambling on the outside chance that most of the scientists are wrong. . . . Let's not take that gamble with the future of our children and grandchildren. . . . Let's enjoy science fiction like Jurassic Park . . . but let's base our decisions on scientific facts.

Senator Hillary Clinton of New York was even more forthright: "Mr. Crichton's critiques of climate change science appear in a work of fiction. His views have not been peer reviewed. They do not appear in any scientific journal." All good points. She then proceeded to use the balance of her time to deliver a scathing critique of the Bush administration's war on science:

This Administration has taken politicization of science to new levels. . . . When scientific knowledge has been found to be in conflict with its political goals, the administration has often manipulated the process through which science enters into its decisions. This has been done by placing people who are professionally unqualified or who have clear conflicts of interest in official posts and on scientific advisory committees; by disbanding existing advisory committees; by censoring and suppressing reports by the government's own scientists; and by simply not seeking independent scientific advice. Other administrations have, on occasion, engaged in such practices, but not so systematically nor on so wide a front. Furthermore, in advocating policies that are not scientifically sound, the administration has sometimes misrepresented scientific knowledge and misled the public about the implications of its policies.

Senator Clinton noted that this was not just her opinion, but the opinion of hundreds of prominent scientists. In fact, her statement directly quoted from a letter (February 18, 2004) that criticized the Administration's misuse of science, signed by a large number of scientists, including Nobel laureates, National Medal of Science recipients, former senior advisers to administrations of both parties, numerous members of the National Academy of Sciences, and other well-known researchers.[6] According to the Union of Concerned Scientists, over the following four years, 15,000 U.S. scientists added their names in support of this letter, to restore scientific integrity in policymaking.

As I listened to Senator Clinton, I wondered who was peddling the real science fiction. Sadly, and alarmingly for the nation at large, fiction about science was being dispensed from on high, like Prozac for the masses, deadening people's senses to ensure that nobody would care too much about many critical environmental issues. Hearings like this one only served to reinforce that perspective. Perhaps having a medical doctor present was appropriate after all.

2
A Letter from Congress

The Pilgrim Trail from Le Puy en Velay in France to Santiago de Compostela in Spain passes through some of the most beautiful countryside in Europe. My wife, Jane, and I had been walking along this route for the past few years, and in the summer of 2005 we went farther, finally crossing the Pyrenees, leaving the quiet villages and lush fields of southwestern France for the drier and more open landscape of northern Spain. I had initially been reluctant to go on this walk; I was too busy and not at all interested in retracing the steps of medieval pilgrims. Surely I would be bored stiff. As it turned out, nothing could have been further from the truth. This trip was relaxing and inspirational, just the break I needed from a hectic schedule of teaching, research, proposal writing, and scientific meetings. By the time we arrived in Bilbao, ready for the flight back home, I was feeling no stress; life was good.

"I'll just take a look at my e-mail," I said to Jane, leaving her to soak in a hot bath. I find that e-mail is like a powerful magnet: the closer you come to an Internet connection, the stronger the attraction. When we finally arrived at a hotel with high-speed Internet, the temptation to log on was irresistible. But as I watched hundreds of messages flash by, it was clear that something unusual had happened. Dozens of them referred to a letter, Congress, inquiries, and testimony. All of this merely washed over me. I could not download any attachments, so I wasn't able to grasp the central issue. After ten minutes it dawned on me that checking my e-mail had been a mistake. I wasn't ready to go back to work just yet. I could postpone the inevitable for another day or two.

"Anything interesting?" asked Jane when I returned to our room.

"Well, something's going on, but I can't figure it out. There were dozens of messages back and forth about some letters. I guess I'll find out what it's all about when we get home. Let's go and find a place to eat."

Back in Amherst a day or two later, it didn't take long to figure out what the fuss was all about. There was indeed a letter—or more specifically a fax—that had been unceremoniously buried in a large box containing all the accumulated mail of the previous few weeks. But this was no ordinary letter. It came from the House of Representatives, signed by the chairman of the House Energy Committee, Congressman Joe Barton of Texas and, more ominously, Congressman Ed Whitfield of Kentucky, chairman of the Subcommittee on Oversight and Investigations.

In fact, as I soon learned, a similar letter had been sent to my "hockey stick" coauthors, Mike Mann and Malcolm Hughes, as well as to the director of the National Science Foundation, Arden Bement, and Rajendra Pachauri, chairman of the IPCC. The opening sentence set the tone for the rest of the letter: "Questions have been raised, according to a February 14, 2005 article in *The Wall Street Journal*, about the significance of methodological flaws and data errors in studies you co-authored of the historical record of temperatures and climate change."[1]

Not a good start. I quickly scanned down the page:

We understand from the February 14 Journal and . . . other reports that researchers have failed to replicate the findings of these studies, in part because of problems with the underlying data and the calculations used to reach the conclusions. Questions have also been raised concerning the sharing and dissemination of the data and methods used to perform the studies. . . . As you know, sharing data and research results is a basic tenet of open scientific inquiry, providing a means to judge the reliability of scientific claims. The ability to replicate a study, as the National Research Council has noted, is typically the gold standard by which the reliability of claims is judged. Given the questions reported about data access surrounding these studies, we also seek to learn whether obligations concerning the sharing of information developed or disseminated with federal support have been appropriately met.

A list of demands then followed, "pursuant to Rules X and XI of the U.S. House of Representatives." These demands included:

- A list of all financial support you have received related to your research, including, but not limited to, all private, state, and federal assistance, grants, contracts (including subgrants or subcontracts), or other financial awards or honoraria

- Regarding all such work involving federal grants or funding support under which you were a recipient of funding or principal investigator, provide all agreements relating to those underlying grants or funding, including, but not limited to, any provisions, adjustments, or exceptions made in the agreements relating to the dissemination and sharing of research results

- Provide the location of all data archives relating to each published study for which you were an author or co-author and indicate: (a) whether this information contains all the specific data you used and calculations you performed, including such supporting documentation as computer source code, validation information, and other ancillary information, necessary for full evaluation and application of the data, particularly for another party to replicate your research results; (b) when this information was available to researchers; (c) where and when you first identified the location of this information; (d) what modifications, if any, you have made to this information since publication of the respective study; and (e) if necessary information is not fully available, provide a detailed narrative description of the steps somebody must take to acquire the necessary information to replicate your study results or assess the quality of the proxy data you used.

- Regarding study data and related information that is not publicly archived, what requests have you or your co-authors received for data relating to the climate change studies, what was your response, and why?

- Explain in detail your work for and on behalf of the Intergovernmental Panel on Climate Change, including, but not limited to: (a) your role in the Third Assessment Report; (b) the process for review of studies and other information, including the dates of key meetings, upon which you worked during the TAR writing and review process; (c) the steps taken by you, reviewers, and lead authors to ensure the data underlying the studies forming the basis for key findings of the report were sound and accurate; (d) requests you received for revisions to your written contribution; and (e) the identity of the people who wrote and reviewed the historical temperature-record portions of the report, particularly Section 2.3, "Is the Recent Warming Unusual?"

The letter was dated June 23. They wanted a response by July 11. It was already July 3. I couldn't possibly get all this information together in the time available. My head was spinning. What brought this on? Who are these people? How can I assemble all the records they need? What if I don't get it done in time? Why is the House of Representatives involved in this? Will I be subpoenaed to testify? Have I done something illegal? Can I be arrested?

I closed my eyes. Oh, to be back on the Pilgrim Trail. Who needs this? But then I started to think more rationally. What is all of this really about? I looked at the letter again, more carefully this time. "Questions have been raised, according to a February 14, 2005 article in The Wall Street Journal." The *Wall Street Journal*? That bastion of scientific authority? No credible scientist gives any credence to that Republican broadsheet. I read on. "In recent peer-reviewed articles in Science, Geophysical Research Letters, and Energy & Environment, researchers question the results of this work. As these researchers find, based on the available information, the conclusions concerning temperature histories—and hence whether warming in the 20th century is actually unprecedented—cannot be supported by the Mann et al. studies." Okay. Now I got the picture.

There had indeed been criticism of our paper, not an uncommon event in the world of scientific research, and we had pointed out that most of the criticisms were erroneous. In fact, the journal *Nature* had evaluated our response to the criticisms leveled at our work and decided that there was no merit in publishing the discussion. The critics had then gone elsewhere to promote their argument, to publications that did not give us the opportunity to rebut their points, and so they managed to get their erroneous ideas into print.

But this was academic trivia. Surely it didn't rise to the level of a congressional inquiry. How could these criticisms have triggered the ire of the Subcommittee on Oversight and Investigations of the House Energy Committee? Didn't they have more important things to do? But there was more:

> The authors McIntyre and McKitrick (*Energy & Environment*, Vol. 16, No. 1, 2005) report a number of errors and omissions in Mann et al., 1998. Provide a detailed narrative explanation of these alleged errors and how these may affect the underlying conclusions of the work, including, but not limited to answers to the following questions:
>
> a. Did you run calculations without the bristlecone pine series referenced in the article and, if so, what was the result?
>
> b. Did you or your co-authors calculate temperature reconstructions using the referenced "archived Gaspé tree ring data," and what were the results?
>
> c. Did you calculate the R2 statistic for the temperature reconstruction, particularly for the 15th Century proxy record calculations and what were the results?
>
> d. What validation statistics did you calculate for the reconstruction prior to 1820, and what were the results?
>
> e. How did you choose particular proxies and proxy series?

Clearly, the House Energy Committee had no interest in Gaspé tree ring data or validation statistics. They probably wouldn't even recognize a validation statistic if it bit them in the rear end. It seemed this was a put-up job. Some special interest had used its political connections to push an agenda, and Joe Barton and his cronies had been all too willing to fall in line.

What was the real issue? Our research had shown that temperatures in recent decades were warmer than at any time in the last thousand years. This was not what the energy companies and anti–global warming activists wanted to hear. And the fact that our study had been prominently reported by the IPCC made the IPCC a threat to the special interests as well. That's why a similar letter had been sent to Pachauri.

All of this occurred to me in a flash. This was not a legitimate inquiry; it was politics, impure but simple. We had unwittingly stumbled into a minefield where the players were the energy companies with infinite financial resources, along with Washington lobbying fronts and political hacks in Congress. Energy company representatives wouldn't stand still for anything that might conceivably impact their bottom line. I imagined a smoke-filled room, full of cigar-chewing oilmen deciding that this global warming business was getting out of hand. Time to make sure that Congress didn't cave in to the damn environmentalists! "Get Joe on the phone and tell him to get on this." I got the picture, and it wasn't pretty.

It wasn't long before my own phone rang. It was Mike Mann, first author of the offending articles. Mike was two weeks ahead of me on this business, and he was charged up. Mike has always had a somewhat conspiratorial worldview, recognizing that some people may act for their own devious purposes, not always without malice aforethought. This is not the way I view the world, but as it turned out, he is often right about such matters. In any case, this inquiry played right into Mike's mindset.

"You see what this is all about, don't you? It's Barton being manipulated by the special interests. They don't give a damn about our published research. They just want to tie us up in knots and prevent us from getting any more work done." Yes, this was becoming clear to me. If I responded fully to everything the congressmen had requested, I'd be unproductive for a long time. Perhaps we could get them to give us longer to respond, or reduce the list of demands. After all, they seemed to be focused mainly on our recent research papers, yet their demands covered my entire career. I probably could not even figure out what financial support I had received over the past thirty years, "including, but not limited to, all private, state, and federal assistance, grants, contracts (including subgrants or subcontracts), or other financial awards or honoraria." Were they kidding? Honoraria? These are private transactions in exchange for giving talks, writing reviews of book drafts, evaluating proposals. What had that got to do with Congress?

Mike laughed. "Exactly. They have really overreached. They've gone way too far, and we don't want to give them any opportunity to seem more reasonable. Don't talk to them. I think you will see quite a negative reaction before too long." Mike was clearly fully engaged in this issue. He had been in touch with a public interest lawyer, who was advising him to take quite an aggressive stance against this inquiry.

Shortly after I spoke to Mike, Malcolm Hughes called from Tucson. He was clearly far less optimistic about where this situation might lead. Unlike Mike, who seemed to relish the battle ahead, Malcolm saw a longer war, waged over access to data and samples, with potentially endless requests for documentation of field notes, laboratory procedures, and decisions about the inclusion or exclusion of data in a particular study. This is how our careers will end, he seemed to be saying, awash in litigation and bickering.

All this talk of lawyers got me worried. I recalled a brief conversation with Myron Ebell, attack dog of the Competitive Enterprise Institute, a Washington-based front for big business interests and a vocal opponent of the notion that global warming has anything to do with human activity. "We want to see you in court," he had told me, which at the time seemed a pretty odd statement to make.[2] See me in court? About what? Did he want to put global warming on trial, as in the famous Kansas evolution "Monkey Trial"? I just figured he was a nutcase and walked away from him, but now I began to think that maybe the gang at the Competitive Enterprise Institute had figured out a way to claim that we had broken some law or other. It wouldn't matter to them whether we had or not, as they had infinite resources behind them. But it could be financially ruinous for us.

I decided I'd better contact a lawyer and get some advice. Not sure how to proceed, I e-mailed the chancellor of the University of Massachusetts, John Lombardi. I received his response very early the next day. He never sleeps much, and it was sent around 4 am. He got the bigger picture immediately. This was far more than an attack on Ray Bradley and colleagues. It was an assault on scientific inquiry. Members of Congress had no business sifting through arcane scientific arguments about validation statistics. If they made inroads on this matter, where would it end? There were plenty of Congressmen who would be happy to push their Luddite agendas on evolution, stem cell research, and who knows what else. This threat had to be taken seriously, and Lombardi was all over it.

As soon as Lombardi arrived in his office that morning, he got the university counsel, Brian Burke, on the phone and directed him to review the House letter and advise me how to respond. This was comforting; although I was beginning to formulate a response in my mind, it was reassuring to talk to a lawyer and get his perspective, especially knowing that the university would stand behind me if things got nasty. Brian's advice was helpful. "This isn't a subpoena," he said, "and your job is to make sure that it doesn't turn into one. So give them just enough, without acquiescing to all of their demands."

Obviously this was going to be a difficult balancing act. But as I reflected on the situation we were in, it just seemed ludicrous, like a bizarre scene from a Fellini movie. Except that this was not at all entertaining, and the stakes were clearly very high.

Unwittingly, we had stumbled into a political storm that had been brewing ever since Al Gore signed the draft Kyoto Treaty in Japan. From that moment on, the stakes over human-induced climate change had been raised significantly, and there were many powerful interested parties who were determined to slap down any notion that might capture the public's attention. Ironically, because of their actions, the results of our little study were headline news—discussed on CNN and the BBC and reproduced in *Time* magazine and in the pages of *Newsweek*. And suddenly it had a catchy name: the hockey stick. We were a brand, and it was being marketed far and wide.

3

The Hockey Stick Controversy

What had we done to enrage the House Energy Committee and its Subcommittee on Oversight and Investigations? What conclusions had we come to that had triggered their fury? In a nutshell, we had concluded that temperatures were warmer in the 1990s than at any time in the previous thousand years, and we suggested that this was due to human actions. No state secrets were revealed; no top secret memos were leaked; no covert operation was launched. We carried out our research, published it after a rigorous review process, and moved on to the next topic. But suddenly an icy hand had reached out and put a chill on our activities. How did we get to this point?

I had been working for years to put together a picture of past climate variations, first using temperature and rainfall measurements from around the world, but then using climate proxies—natural archives such as tree rings, ice cores, corals, and sediments—to extend the record further back in time. Much of this work was collaborative, performed with colleagues at the University of East Anglia in Norwich, England, and with specialists at the Laboratory of Tree-Ring Research at the University of Arizona in Tucson. This research began long before it was apparent that global warming, due to greenhouse gases from the burning of fossil fuels, was becoming a problem. In fact, 50 percent of "anthropogenic carbon dioxide" (the excess carbon dioxide that has entered the atmosphere since the Industrial Revolution) has been produced since I was in graduate school, which makes me recognize how fast we are burning fossil fuel, as well as how depressingly old I'm getting.

My research was aimed at understanding how climate had changed over time and what factors may have played a role, with natural effects such as explosive volcanic eruptions and changes in solar activity high on the list of likely suspects. I was never a fanatical proponent of the theory of human-induced global warming, but as time went on and the records kept being broken, it became increasingly obvious that we were entering a new era in which human activity was affecting global temperatures.

Nowadays, the links between human activity and global warming are apparent everywhere you look; but when changes develop slowly, year by year, it's hard at first to separate the human-driven effects from natural background "noise." As we've seen, there is a vocal chorus of opposition to these ideas. More often than not, though, critics are arguing over trivial details that have absolutely no effect on the big picture. Like it or not, the warming really is happening and it really is global. Furthermore, we shouldn't be too surprised by this. We have burned so much fossil fuel that the fundamental energy balance of the atmosphere has been altered within the short space of just 150 years. We have no evidence for any other such rapid change on earth in the entire history of *Homo sapiens*.

As discussed earlier, simple physics indicates what will happen if carbon dioxide levels rise: it will get warmer. The difficulty is in figuring out what other factors might also change so as to reduce (or perhaps amplify) this effect. To work that out, we use computer models of the atmospheric and oceanic circulation, and these provide a picture of what we might expect to occur as carbon dioxide levels continue to rise. The advantage of these models is that they give us not just the overall global temperature but the geographical and seasonal patterns of expected changes, and also show how the changes vary at progressively higher levels in the atmosphere. Furthermore, the models provide the same information about other potentially important factors that might affect the climate—what are called "forcing factors," as they force changes in the climate. For example, we can use the models to simulate the expected patterns of change that would result from an increase in the amount of energy the earth receives from the sun. Some have argued that global warming is entirely due to an increase in solar activity; but the pattern of change we would expect from such an effect, the diagnostic "solar fingerprint," is entirely different from that which would result from an increase in greenhouse gases. Thus the models provide multiple ways to see if the changes we are observing fit the expected pattern for greenhouse gases or if they are related to other forcing factors. Indeed, this complex fingerprint of expected change due to an increase in greenhouse gases is pretty much what we are seeing play out: warming is greatest in the winter months, at high latitudes, and in the high mountains of the Tropics, but far up in the stratosphere, temperatures are cooling.

Scientifically, the noose of evidence that the burning of fossil fuel is affecting global climate is slowly but surely tightening around the necks of the critics, the diehard contrarians who desperately cling to ever more hopeless arguments that carbon dioxide has nothing to do with the observed changes, and that everything is just part of a grand natural cycle. After all, it was warm in the past, right? What about the dinosaurs? Too far back in time? Well, what about the Vikings? Yes, the Vikings! They sailed away to Greenland. They even called it "green," so it must have been warmer then. And then they died out, so that proves it must have become colder, and now it's warm again. *Voilà!* Global warming explained.

This tired old pathetic argument has been rolled out at so many public meetings, and reproduced in so many blogs, that it has taken on a life of its own. If you repeat something often enough, the thinking goes, it surely must be true. Thankfully this is not so. In truth, temperatures in northern Europe were at times mild during the days of the Vikings, and they were much colder a few centuries later. But this had nothing to do with the Vikings setting off from Scandinavia, which happened for a host of political, social, and religious reasons. More important, there is no clear evidence that the world was warmer than usual *everywhere* in medieval times. In fact, some places were certainly cooler, so when you average it out, there really isn't any compelling evidence for *global* warming in the past like the warming we have seen in recent decades, when temperatures have risen just about everywhere.

There is, however, a more general point that the denialists cling to. How do we know that there hasn't been global warming of the kind we've seen over the past century at some point in the past? Weren't there times when temperatures rose just as much as they did in the twentieth century, and if so, what caused such an increase?

These are the sorts of questions we climatologists do find of interest, not simply for the sake of understanding global warming, but more generally as a way to see how the climate system has changed over time, what factors might have brought about those changes, and how changes at a global scale were related to changes in different geographical regions. These were the topics that Mike Mann and I agreed to work on together. We recognized that to do this right, we'd also need the advice and expertise of Malcolm Hughes, a specialist in tree ring research, to help us expand the database on which we could build a global picture of past changes in climate. We then began to formulate a research plan to reconstruct how global or hemispheric temperatures had changed over the last thousand years. Little did we know what we were getting into.

Mike Mann actually grew up in Amherst, where I have lived for over thirty-five years while teaching at the University of Massachusetts. In fact, he went to high school with my son Stephen. Stephen recalls Mike as a popular science class partner because he always got the assignments right. Then he went off to the University of California, Berkeley, to study physics and applied mathematics, before returning to New England to pursue an M.S. in physics and a Ph.D. in geophysics at Yale. I never knew him in those days, but quite by chance I met his father, a mathematics professor at the university, at a wine tasting held at a friend's house one evening. As we swirled our cabernet sauvignon and gradually sank lower into comfortable chairs, we struck up a conversation about what we did for a living. When I explained my interest in climate change, Mike's dad remarked that this sounded like what his son was interested in, and perhaps we should get together next time he was home. So we did, and it turned out that Mike was doing exactly what was of interest to me. He and I soon began to think about how we might collaborate in our research. Mike had heard of a rather competitive Department of Energy fellowship (the Alexander Hollaender Distinguished Post-doctoral Fellowship), and if I would agree to host him, he would write a proposal to come to the University of Massachusetts so we could work together on issues of mutual interest. Despite a lot of competition, Mike's proposal was successful. And so it came to pass that he eventually came full circle and returned to Amherst, in no small part thanks to a very good claret.

Mike Mann is a clever guy. His knowledge of statistical methods and his understanding of the climate system are exceptional. While in my climate research group, he was full of ideas. Soon he proposed a fairly innovative new approach to reconstructing how temperatures had changed in the past.

Because the record of temperatures made by thermometers is so limited in time, if we want to get a picture of how global or hemispheric temperatures varied before about 1850, we have to rely on "climate proxies." Studies of these proxies fall under the heading of paleoclimatology. Proxies are simply natural phenomena that have captured in some way a record of past climate. The simplest examples are tree rings. In many regions where tree growth is limited by temperature (for example, at the northern limit of trees in the sub-Arctic, or high on mountain ranges at the tree line), year-to-year variations in temperature greatly affect the ability of the tree to produce cellulose and thus create new wood. By extracting a thin core of wood from a tree, we might see patterns of narrow and wide rings, which signify cool or warm years in the past. By counting the rings and measuring the width variations, we gain a "proxy" record that represents past temperature changes. Of course, there are many other factors that might affect tree growth, so the tree ring specialists who carry out this kind of research are trained to select trees in which the climate "signal" is likely to be maximized and other, non-climatic factors are likely to be minimized. They also take many core samples from several trees in each area to reduce the "noise" in the records and strengthen the climate signal that is of interest. Literally thousands of such samples have been collected in this way, from sites all over the world.

Another method of obtaining the past history of temperature is to look at the ratio of oxygen isotopes (atoms of oxygen with different atomic mass) in water molecules found in ice cores from polar regions. Snow that falls on the high ice sheets of the Arctic and Antarctica (and on ice caps in high mountains around the world) may not melt before it is buried by a new snowfall. The snow builds up in layers, and by coring through the ice sheets, you can recover a record of past snow accumulation that represents the history of precipitation from the atmosphere onto that location going back many years. The ratio of oxygen isotopes ($^{16}O/^{18}O$) is related to the temperature of the air when the snow originally fell from the clouds. This can also provide paleoclimatologists with a record of past air temperature.

Other archives, far from polar regions, can provide additional information to help in building a global picture. In many parts of the tropical oceans, large corals can be found, growing as domes of calcium carbonate, layer upon layer, year by year. By drilling through these domes, paleoclimatologists can extract cores that reveal yearly growth increments, often extending back several centuries. The organisms that produce the calcium carbonate are sensitive to the temperature of the ocean water, and the chemical composition of the carbonate deposited (specifically the ratio of the two main oxygen isotopes in the carbonate) varies with the temperature of the surrounding water, so in effect the corals can provide a history of the local sea surface temperature.

Finally, in some areas, sediments from the bottom of lakes can be extracted and the material studied to obtain a temperature history for the lake region. These findings might be related to the thickness of each sediment layer, which reflects the amount of sediment transported to the lake each year. The amount of sediment may vary, depending on the amount of rainfall, or the rate at which winter snowfall melted as temperatures warmed up in the spring. Or the sediments might be studied for their biological content, which may reflect temperature during the growing season. For example, in some lakes the type and amount of diatoms growing in the lake each year are dependent on the duration of ice cover on the lake. If the winter is mild, the ice may form only briefly (or be thin), resulting in a long growing season for the diatoms, and this will be reflected in the sediments. In short, there is a wide variety of climate proxies that we can use to build up a picture of how temperatures varied in the past. Of course, they are not direct measurements such as those we can obtain from thermometers; but nevertheless they can be remarkably good at revealing temperature changes over time.

In contrast to most previous studies, Mike Mann's idea was to use a very large set of data from all over the world, and to apply powerful statistical techniques to extract from these data information about temperature patterns. Rather than just look at a few records and relate them to local temperature changes, we would examine extensive networks of data, recognizing that even local temperature changes are related to the larger-scale patterns in the atmosphere, so that variations in one region might very well be linked to conditions far from that location. For example, we might have a proxy record from southern California that indicated a period of severe drought. In that part of the world, drought is generally caused by changes in atmospheric circulation brought about by shifts in surface temperatures in the Pacific Ocean; specifically, droughts are associated with the cool-water equivalent of an El Niño, known as La Niña. La Niñas and El Niños involve oscillations of water masses in the equatorial Pacific, and these have associated changes in atmospheric circulation. A La Niña–related drought record thus carries with it far more information than just the fact that there was below-normal rainfall in southern California. It is connected to a global-scale phenomenon that has links (teleconnections) to weather patterns all over the world.

By using records such as these, as well as records more directly related to local temperature, we looked for patterns in the proxy data that matched patterns in the recorded temperature records for the period when the two sets of records overlapped (1900–1980). This provided a statistical key, or calibration, which we could then apply to the proxy records in the period before instrumental records began, to unlock the patterns of temperature change that had occurred year by year during that time. Nobody had done this kind of research before in quite the same way; it was pioneering work, and we inched our way forward as we tried to decide on the optimum procedures. Mike was in charge of the statistical approach, writing the computer code and calculating the uncertainties as we pushed the reconstruction back in time, while Malcolm and I selected the different records that would be used in the analysis.

After a lot of effort we finally produced a grand reconstruction, extending all the way back to AD 1400. This graph showed that average temperatures across the Northern Hemisphere had been about 0.5°C lower than in the twentieth century for most of the preceding period, but there was a sharp rise in temperatures beginning in the late nineteenth century. The coldest fifty-year period before 1900 was in the fifteenth century (1453–1502). We also showed that most of the wiggles in the record could be explained by two natural factors: explosive volcanic eruptions and changes in solar activity. Or at least these factors explained almost everything until the start of the twentieth century, when temperatures rose without any corresponding change in either volcanic or solar activity. The only explanation we could find for temperature change in the twentieth century was that increasing levels of greenhouse gases had driven temperatures upward, decoupling the hemisphere's temperature from the underlying natural factors that had been in play throughout all the previous centuries.

This was a really "cool" result, and definitely worthy of a submission to the most prestigious science journal, *Nature*. Mike wrote it up, and we submitted it in 1997. After some time, the paper came back with quite extensive comments and questions from reviewers. They wanted more details on this and that; they couldn't follow such and such an argument, and they would like to see more detail here, less emphasis there, and so on. Never discouraged, Mike set about responding to each comment. Some were valid and important: we had not explained a point carefully enough, or we had overlooked a paper published just recently that should have been cited. Others were less relevant and easily dismissed. The normal process of submission, review, and response played out over the ensuing months, with reviewers getting a chance for a second look at the revised manuscript before it was finally accepted and published, almost a year after the original paper was drafted.

The paper made quite a splash. I happened to be in London at a conference on April 22, 1998, the day it was published,[1] and found myself being interviewed on the 7 pm national evening news. I thought I had been asked to appear simply to comment on the paper and its main conclusion—that human activities had started to overtake natural factors in driving up temperatures across the Northern Hemisphere—but when I arrived in the "greenroom" to be daubed with makeup, I was surprised to find Nigel Calder sitting there as well. I had known him only from his popular science books on astronomy and the weather, so I was quite pleased to meet him. As we were called in to the studio, he leaned over to me and said, "I'm going to destroy you."

It did not take long to see where the discussion was going. The interviewer was determined to put the *Nature* paper on trial, and Calder's job was to trash it. I found this amusing, and in the end I don't think he scored many points. But perhaps I should have taken heed even then: the paper we had just published was not cool at all. It was hot! And things would only get hotter as time went by.

We were not the only scientists working on ways of reconstructing temperature changes over the past millennium. There was a lot of interest in the subject, because when people looked at the record of measured temperature over the last century, with its remarkable increase in recent years, the obvious question was: "How unusual is that. Is it just part of a natural cycle?" We were particularly motivated to try and push our perspective on temperatures of the past century as far back in time as we could in order to answer this question. In the *Nature* paper we had gone as far back as 1400, but we stopped at that point because we had only a slim network of records for earlier periods. After the *Nature* paper had been published, however, we decided that it was just possible to produce a reconstruction that covered the last thousand years, though the uncertainties would be quite large. By applying the same techniques to the few records that extended back that far, we extended the original temperature estimates back to AD 1000. To make sure that people were informed about the uncertainties that such an effort involved, we titled the paper "Northern Hemisphere Temperatures during the Past Millen-

CHAPTER THREE

nium: Inferences, Uncertainties, and Limitations." The article appeared in the journal *Geophysical Research Letters* on March 15, 1999.[2] In the abstract we stated, "We focus not just on the reconstruction, but the uncertainties therein, and important caveats," and much of the paper involved a discussion of how difficult it was to be sure about the earliest part of the temperature record. In other words, we hung red flags all over the place—in the title, the abstract, and the text—so that readers could appreciate the tentative nature of the reconstruction.

Even with those uncertainties, it appeared that the most recent decade (the 1990s) was the warmest of the entire thousand-year period, even when we took into account the potentially large errors associated with our estimates of temperature far back in time (figure 1). The 1990s were 0.34°C warmer than the next-warmest decade, way back in the twelfth century (1166–1175). Furthermore, it was clear from all instrumental records that 1998 was the warmest year of the twentieth century, and so logically, 1998 was likely to have been the warmest year of the millennium; in fact, the next-warmest year in our temperature reconstruction, 1249, was more than 0.5°C cooler.

We expected that this conclusion would be newsworthy, but we were not at all prepared for the media frenzy it produced, or for the subsequent inquisition led by Congressmen Barton and Whitfield. One lesson we learned from this was that once you let the genie out of the bottle—and in our case, the genie was the thousand-year temperature graph—you can't get it back in no matter how many times you point out the caveats and cautionary statements that originally accompanied the paper ("inferences, uncertainties, and limitations"). Caveats do not lend themselves to one-column articles in *Newsweek* or *USA Today,* or sound bites on CNN. Nobody cared about the details, and the headlines quickly eclipsed reality. Hence the graph, dubbed "the hockey stick" by Jerry Mahlman (director of the Geophysical Fluid Dynamics Lab in Princeton, New Jersey), soon became an icon, a powerful symbol of the human impact on climate. Consequently it became a magnet for all those forces that sought to dismiss and trivialize the idea of human-induced climate change.

This situation was certainly made more acute by the fact that the

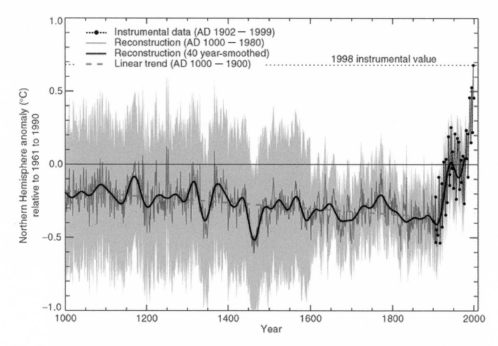

FIGURE 1. The "hockey stick" graph, showing our estimate of average tempera-tures across the Northern Hemisphere since AD 1000, based on temperature-sensitive "proxies" such as tree rings, banded corals, ice cores, and sedimentary records. All values are shown in relation to the average for 1961–1990 (the zero line on the graph). The dark gray shows the individual yearly values and the lighter gray shows the uncertainty of those estimates. The heavy black line is a smoothed version of the individual yearly values. The dotted line, which overlaps with the reconstructed values, shows the actual measured temperatures from instrumental records across the Northern Hemisphere. The highest annual average value was in 1998, and that exceeded all previous values, even taking into account the calcu-lated uncertainties, as shown by the dashed line at the top.

Source: Michael Mann, Raymond S. Bradley, and Malcolm K. Hughes, "Northern Hemisphere Temperatures during the Past Millennium: Inferences, Uncertainties, and Limitations," Geophysical Research Letters 26 (1999): 759–62.

CHAPTER THREE

IPCC highlighted the hockey stick graph in its 2001 report. I explain more about the IPCC later (in chapter 4), but for now it is important to know that the scientists involved in producing the IPCC reports are charged with simply reviewing published results. The "IPCC panel" (which is really made up of many teams of scientists from all over the world) looks at whatever was published over the period since the previous IPCC report and provides a synthesis or summary of that work. Thus the IPCC acts as a sort of monitor of progress in the field of climate change, and its reports, issued every four years, are very influential in providing governments with an assessment of what is known at that moment in time.

Apart from the weighty tomes that it produces, full of the nuts and bolts of climate science, carbon dioxide emissions data, and assessments of how global warming may affect different economic sectors from forestry to fisheries, the IPCC also produces a "Summary for Policymakers," who are (rightly) assumed to be too busy to read the whole thing. A few figures are included in this summary to highlight some of the important findings that have appeared in print over the previous few years. In the 2001 report, one of those figures was the hockey stick graph from *Geophysical Research Letters,* but stripped of all the text we had carefully constructed about limitations and uncertainties. Not that it would have made a lot of difference; the graph was quickly selected by almost every major media outlet as the essence of the IPCC report. It was simple and eye-catching, just what the media were looking for to represent this weighty and (to many) largely incomprehensible report. Consequently, the obscure hockey stick very quickly became a media symbol of the global warming problem, and thus a target for people intent on destroying the credibility of those in the scientific establishment who had labored long and hard to produce a fair and balanced assessment. To the critics, sowing doubt about the hockey stick was akin to casting doubt on global warming in general, and thus a way to block legislative action on limiting carbon emissions from fossil fuel. As ludicrous as this reasoning may seem, it nevertheless was the perverse logic that precipitated the letters and the associated hearings, led by Chief Inquisitors Inhofe and Barton.

As it happened, Mike Mann was one of eight "lead authors" of the 2001 IPCC chapter dealing with "observed climate variability and change." This chapter had two senior authors, one from the UK and one from the United States, and 140 "contributing authors," of whom I was one and Malcolm Hughes was another. To a conspiracy theorist (of which there are many on the subject of global warming), the fact that the hockey stick graph was included was clearly the result of personal bias and prejudicial pressure applied to our 147 collaborators by the three of us. Just to put that absurd idea in perspective, ours was only one of about forty figures in the chapter. More than seven hundred research papers were mentioned in the chapter, of which only fifteen were by me, Hughes, and/or Mann. It seems farfetched, to say the least, to claim that the chapter was biased by our participation in writing it. But the critics did not stop even at that extremely tenuous conclusion. They claimed that the entire volume—all 881 pages of it—was contaminated by inclusion of the toxic hockey stick, and consequently all of its conclusions had to be dismissed. Apparently everything we thought we knew about greenhouse gases and global warming rested on the hockey stick, and if it could be shown that the stick was broken, nobody should take the matter seriously. It was all a pack of lies, or as Senator Inhofe was fond of saying, just a hoax.

After the IPCC report was published, criticisms of our work soon appeared. Critical evaluations of published research are not unexpected. This is how science progresses: somebody publishes a research paper, then others examine the method, the data used, or the conclusions reached, and point out inconsistencies in the approach, or new and better data that could have been used. Or they might suggest another, entirely different approach, which could lead to different conclusions. We considered our temperature reconstruction a "working hypothesis," based on the best data we could assemble at the time and what we, at least, judged to be a well-thought-out method of analysis. If others could show that there was a better approach, it might lead to a new "working hypothesis," and so the science would slowly move forward. All this would normally take place within the framework of what are known as "peer-reviewed journals."

In science, when a paper is submitted for publication, it is sent to two or more reviewers, who read it and advise the editor of the journal whether it is worthy of publication. Generally these reviews are anonymous, so the reviewer can give his or her unvarnished commentary and an honest assessment of whether the paper is ready for prime time. More often than not, the reviewers advise the editor that something has to be fixed before publication; perhaps a section needs clarification, or further analysis is requested. The process is orderly and has proved its worth over the years. If an editor receives a paper that directly criticizes something that has already been published, the authors whose work is being criticized are invited to respond, and the editor will then decide if the discussion is worth publishing or if the points being raised are trivial or simply wrong. If the editor accepts the criticism as valid, both the critique and the response are published together so readers can see the extent of the disagreement and make up their own minds. After all, it is only common courtesy to allow somebody being tapped on the shoulder (so to speak) to turn and confront the person who is demanding attention.

In view of this, we were rather surprised when a criticism of our *Nature* paper appeared in June 2003, in an obscure British journal, *Energy & Environment.* In fact it was a lot more than a criticism. The Canadian authors, Stephen McIntyre and Ross McKitrick, claimed to have "audited" our work and then "corrected it," concluding that the hockey stick was nothing more than "an artefact of poor data handling, obsolete data and incorrect calculation of principal components."[3] Now to most of us, the word "audit" has quite a specific meaning. In an audit, an accountant carefully checks all the transactions and makes sure that nobody has been cooking the books. It implies that there may have been questionable procedures, requiring careful and credible oversight. That was not the case here. Their analysis of our work was shoddy and inadequate; they did not follow the procedure we had used, and they eliminated certain sets of data. Not surprisingly, therefore, they came up with a different result. Far from "correcting" our work, they bungled the analysis. Had we been given the opportunity to point out where they had gone wrong, it is doubtful that the paper would have been accepted by a typical journal. But *Energy & Environment* was not run like a typical journal. It was merely a vehicle for the editor to promote her own agenda. But for that purpose the article was good enough, and so it appeared in print.

The authors then went on to make a similar pitch to the editors of *Nature,* who asked us to respond. Once they read our response, it was clear that there was no point in publishing the discussion, and so the criticism was rejected. The authors then sent it to *Geophysical Research Letters,* where, surprisingly, the editor did not follow the normal procedure of sending it to us for a response. Without guidance on what was wrong with their analysis, he gave it the green light, and the criticism was thus published without any reply from us.

This may sound like petty bickering to people standing on the sidelines, but for those who were angry about the hockey stick and all that it implied, having a criticism of it appear in a prestigious journal like *Geophysical Research Letters* was a very big deal. Our critics could hold it up as a "peer-reviewed" article, one that had undergone rigorous examination and passed muster. They could then claim that the hockey stick had been proved to be wrong, that it was the result of bad science, poor judgment, and inadequate data, the work of devious and wicked scientists intent on sowing deception and mischief. Or so one might have concluded as the spin machines revved up to a high pitch. All the right-wing propagandists lined up in a chorus to make sure that the Hockey Stick Three would be exposed as charlatans, a bunch of climatological quacks. Leading the charge was the *Wall Street Journal,* whose clientele were happy to read—right on the front page, no less—that the hockey stick was broken and global warming was a hoax, just as Senator Inhofe had been telling them all along. As the train began to gather speed, there to blow the whistle was Congressman Joe Barton, who had probably never heard of *Geophysical Research Letters* until he read about it in the *Wall Street Journal.*

So, what exactly was the controversy? Was our analysis really wrong? So much has been written about this, and so many radio talk shows, TV news shows, and Internet blogs have trumpeted the idea that our analysis cannot be trusted, that the casual bystander could be forgiven for thinking that there really was a problem. For about a day or so after we first read the paper in *Energy & Environment,* even we had a hard time understanding how the authors had obtained a result so different from ours when they claimed to have followed the same procedures. But Detective Mann was on the case and soon discovered the reason: they had not, in fact, repeated our analysis in the same way.

By changing the statistical procedure, these critics had effectively eliminated a sizable set of data from the western United States. It was no surprise, then, that their graph was quite different from ours. Instead of a hockey stick–shaped curve, theirs was more U-shaped, with a strange warm period in the fifteenth century—warm at the beginning and the end. This flew in the face of all that we knew about the fifteenth century, a time when glaciers advanced around the world. A result suggesting that it was warm in that period was obviously unlikely, and should have made the authors wonder what they had done wrong. But the authors were not climatologists, and may not have been familiar with the literature that we all took for granted.

In any case, their audit was simply not a fair test of our work. As we reviewed their paper, however, it became clear that we had not helped those who wished to repeat our analysis because there were some mistakes in the on-line supplementary tables, which listed details about the data sets we had used. We had failed to supply *Nature* with a corrected final list, so there were a few errors in there. This omission had nothing to do with why our critics' result differed from ours, but to make sure it did not confuse anybody else, we published a short, fifteen-line "corrigendum" in *Nature* on July 1, 2004,[4] simply reporting the error and providing a table with the corrected data sources and other details This changed nothing at all in our analysis. As we noted, "None of these errors affect our previously published results."

Corrigenda are not uncommon in scientific journals, and are used to set the record straight; when you work with multiple authors, and written drafts of papers are exchanged back and forth via e-mail, the process can sometimes get confusing. In our case, somewhere along the way the penultimate version of the supplementary data table had been attached to the *Nature* manuscript instead of the final version.

Corrigenda address such problems; they are not retractions of results. But this simple act of correcting the record was enough to set off a chorus of abuse in the blogosphere, and before long, letters appeared in newspapers all over the country, supposedly from outraged citizens. The writers of these letters claimed that the scientific world had been misled by our deliberate deception, and that the editors of *Nature* had "forced" us to apologize and retract the conclusions.

Reading such nonsense was like stepping into the world of Alice in Wonderland. Sometimes the writers even got the name of the journal wrong. They were simply parroting something they had seen, or perhaps had heard on talk radio; they certainly had never actually read what we published in *Nature*. But such is the world of the Internet (and the wild frontier of the airwaves). Anybody with half a brain can set up a website and appear authoritative to those who don't know any better. The same applies to know-it-all loudmouths on the radio. It was for precisely this reason that I joined a group of highly qualified climate scientists to launch the website www.realclimate.org, which tries to debunk stupidities in the press and elsewhere, whenever they appear. This is a big job, as there is an ocean of garbage floating around out there, and trying to sink it all is a full-time job (especially for those of us who already have full-time jobs). Nevertheless, it is important to try to stamp out deliberate misinformation. As a result, I spent much of the summer of 2004 writing responses to letters to the editor that had been written to numerous small-town newspapers throughout the United States in what I can only conclude was a coordinated disinformation campaign. Naturally that took a lot of time which could have been better spent on further climate research. To a large extent the underlying goal of this campaign was probably successful: my time was wasted by the distraction.

In fact, the entire campaign to smear our reputations and sow the seeds of doubt among the public was remarkably effective. It spread like wildfire. The detailed criticisms of the methods we had used were highly technical, involving questions about "centering protocols," Preisendorfer's criteria, and other matters that unfortunately, to all but a small number of people, were incomprehensible. That's not to dismiss them as unimportant; science requires credible arguments and statistical procedures that can be justified. But often the points of contention can be extremely picayune—perhaps of arguable significance to those whose life revolves around arcane issues of statistical procedures, but hardly important in the larger picture of things.

I began to think about this problem one morning as I was making oatmeal for breakfast. For years I had shoved it in the microwave oven: one cup of oatmeal, two cups of water, mix it up and put it in the microwave for one minute and thirty-five seconds. But as I waited this time for it to cook, I casually looked at the instructions on the box. Microwave for one minute and FORTY seconds! For years I had violated all the rules of oatmeal-making. How had I got away with this for so long? I had made oatmeal for others, and nobody had ever called me on it. My prospects for induction into the Oatmeal Makers' Hall of Fame suddenly faded; I had not followed the exact rules. And yet the oatmeal tasted pretty good. Perhaps there was room for a slight deviation from the printed instructions. Maybe, in the end, it really didn't make a damn bit of difference. And that, in a nutshell (or bowl of oatmeal), was about the level of dispute over the statistical procedures we had used in our analysis. Some would have preferred it to cook a bit longer in the statistical microwave, but there weren't any published instructions when we began the research, and it turns out that one minute and thirty-five seconds was a pretty good estimate—and it changed nothing.

Nevertheless, it was clear that the world at large was not going to be persuaded by my oatmeal-making analogy. The only way to settle the matter was to have an independent, authoritative body evaluate all the claims and counterclaims, which is why the National Academy of Sciences got involved. This was not the option that Congressman Barton would necessarily have chosen, as it took matters out of his control, but thankfully, not every Republican was willing to go along with his witch-hunt. The first to step out of line was Congressman Sherwood Boehlert of New York, who blew the entire Barton inquiry out of the water.

The letter from the House Energy Committee sat on my desk for about a week while I pondered how best to respond. It was clear to me that I wouldn't be able to deliver what the congressmen had requested; that would have taken me forever, and my answers would probably have been incomplete anyway. Instead, I began to dissect the request. Sending them my curriculum vitae was easy enough, and since that listed all my publications, it would cover the first point. As for federal support I had received, I could stand on the technical point that, strictly speaking, I had never personally received a dime from the feds, since all the grants I applied for went to the university. I just work there. Hmmm, I began to think, maybe this wasn't going to be so hard after all.

The real breakthrough came one evening as I was watching the news on television. President Bush, who had steadfastly refused even to acknowledge that global warming had occurred and was an arch-opponent of any legislation to reduce energy consumption, was on his way to the G8 summit in Scotland, where Prime Minister Tony Blair planned to make global warming a major issue on the agenda. When Bush stopped in Copenhagen on the way, he was asked by a reporter if he was worried about global warming. His response was stunning: "The surface of the earth is warmer and an increase in greenhouse gases caused by humans is contributing to the problem." This was the first time I had heard him even accept that global warming was a reality, let alone link it to human use of fossil fuels. No doubt he had carefully selected his words: "contributing to the problem" was purposefully ill defined. Was it contributing 1 percent or 99 percent? In any case, a light came on in my mind as this provided me with the perfect introduction to my response. I also decided that the committee members simply did not understand how scientific research was carried out, and they could use a little education on the subject. Certainly scientists had to meet a far higher standard of review than somebody writing an opinion in the *Wall Street Journal*. With this in mind, I launched my response:[5]

Dear Congressman Barton and Congressman Whitfield,

It is good to know that your committee is keenly interested in understanding the basis for President George Bush's recent statement: " . . . the surface of the earth is warmer and an increase in greenhouse gases caused by humans is contributing to the problem." My work has made minor contributions to this issue, which has been the focus of intense international scientific research in recent decades.

There is now very little doubt that President Bush is correct; this is the view held by almost every person who has carefully studied the problem. Greenhouse gas concentrations in the atmosphere are now higher than at any time in at least the last 750,000 years (more than three times the length of time that our species, homo sapiens, has been on earth). It took over 10,000 years for carbon dioxide levels at the end of the last ice age to rise by 100 parts per million (to 280ppmv) but it has taken only ~150 years for concentrations to increase by another 100ppmv. Indeed, about half of that increase has taken place within the last ~40 years, so the rate of increase is unprecedented, and accelerating. At the same time, global temperatures have risen to levels higher than at any time since records began. Our research, and that of many others, suggests that mean temperature in the northern hemisphere is, in fact, higher than at any time in at least the last 1000 years. These conclusions are consistent with theoretical studies dealing with the expected consequences of increased greenhouse gases. That is, theory—supported by modeling studies—predicts that certain changes would be expected if greenhouse gas levels increase as they have done, and these predictions are similar to what we have observed in instrumental records, and in natural archives that are affected by climate changes. It is this very large body of work that led the Inter-Governmental Panel on Climate Change (IPCC) to draw the conclusion in its last report that, "The balance of evidence suggests a discernible influence on global climate." You are quite mistaken in thinking that this conclusion rests largely on the work of Bradley, Hughes or Mann, or on the three of us together. The IPCC Report ("Climate Change 2001: The Scientific Basis," Cambridge University Press) is 881 pages in length. It weighs 5.5 pounds and contains over 200 figures and 80 tables. It would be absurd to think that the weight of its conclusions rests on any one figure or table;

rather it paints a convincing picture in the totality of its science, as noted succinctly in its title.

(In thinking about how ludicrous it was to consider that any one graph in the huge IPCC report was critically important to its overall conclusions, I had decided that the best response was to weigh it on Jane's kitchen scale. It took almost all the weights we had available to balance the enormous IPCC tome.)

You mention that there have been several papers published that disagree with the conclusions of papers published by Mann, Bradley and Hughes. This should come as no surprise. That is the nature of scientific activity. We publish a paper, and others may point out why its conclusions or methods might be wrong. We publish the results of additional studies that may argue with those critics, and provide data that might support or modify our original conclusions. That's normal. Scientific developments generally take place incrementally, one or two steps forward, perhaps one or two back . . . or perhaps a little to the side. But as time goes on, robust results generally become accepted as other studies come to similar conclusions using perhaps different data, different approaches, different starting points. That is where we now stand with respect to our conclusion that the recent warming is unprecedented within the context of (at least) the last 1000 years. Others re-examined our methods and our data and came to the same conclusions that we did. Others have used different data and different methods, but also reached the same conclusion. This scientific approach, following well-established procedures involving the courteous exchange of views, both informally in scientific meetings and formally in the scientific literature, is what moves science forward. It does not move forward through editorials or articles in the *Wall Street Journal* or *USA Today*; it does not advance through ad hominem attacks on individual scientists in the Congress of the United States; it does not move forward through novelists deciding that they can sort the problem out by fleeting references to scientific papers within the pages of fiction. The problem of climate change will be documented through patient and careful analysis, carried out by those with the scientific background necessary to understand the problem.

I then gave some specific responses to their inquiry and closed with a little more education:

> Let me conclude by pointing out that the paper which seems to be the focus of so much of your attention (Mann et al., 1999) was entitled, "Northern Hemisphere temperatures during the past millennium: inferences, uncertainties, and limitations." In fact, a major point of the paper—which both you and others seem to have overlooked—is that we were at pains in this paper to point out the difficulties of drawing conclusions about the climate of the past millennium. We recognize and estimate the uncertainties involved in such paleoclimatic reconstructions. If others choose to ignore those caveats, there's not much we can do about it. Nevertheless, the estimates that we provided have proven to be quite robust and the "working hypothesis" that we presented is now quite well supported by numerous other studies.

By the time I finished writing this letter, I was feeling pretty good about it. I felt I had managed to meet the letter of the law without caving in to their demands. I had tried to steer a course between sneering sarcasm (which no doubt would go right over their heads) and detached cooperation. But I was still nervous about what would happen next. These were powerful people who could create a lot of legal problems for me at the stroke of a pen, so it was nerve-racking to wait for their response.

Meanwhile, Mike and Malcolm had written their own letters, each taking a slightly different tack. Malcolm's was deferential and precise; he peppered it with footnotes, directing them to additional studies that supported our conclusions, and explained that disagreement and debate was a hallmark of scientific research. He might have added that, presumably, a similar process took place in Congress before legislation was passed, but that point would perhaps have been too subtle for these politicians to appreciate.

Like me, Malcolm also tried to disillusion Barton and his co-inquisitors about the extent to which our research had influenced the conclusions of the vast report produced by the IPCC. He closed with a not very genuine invitation: "Should you ever have the opportunity to visit

Tucson, we would be pleased to give you an introduction to the wide and fascinating range of work being done here." He might have added, there's a nice airport in Tucson where you can land your industry-sponsored jet.

Mike's letter had an entirely different tone, beginning with a disclaimer that stated: "This response is submitted without waiving any objection I might have to the Committee's jurisdiction over the subject matter of this inquiry." This was certainly based on advice Mike had received from his legal counsel, and it set him on a much more confrontational course than either Malcolm or I had taken. In fact, by the end of the first paragraph he had given an unequivocal response: "The criticisms your letter cites have been soundly rejected by the scientific community." By then he was on a roll: "The most serious contention in your letter—namely, that my work has not been subject to replication because I have failed to make available the underlying research data— is incorrect. Your letter notes that the National Research Council's 'gold standard' for scientific research is the ability of other scientists to replicate first generation research, and I fully agree. . . . [A]ll of our data and methodologies have been fully disclosed and are available to anyone with a computer and an internet connection."

At this point Mike's response took a different path because the letter he had received from Barton contained a very specific set of questions and requests not included in the letters to me or Malcolm. The congressmen had written, "According to *The Wall Street Journal*, you have declined to release the exact computer code you used to generate your results." The letter then posed a series of questions: "(a) Is that correct? (b) What policy on sharing research and methods do you follow? (c) What is the source of that policy? (d) Provide this exact computer code used to generate your results."

Mike's response went straight to the point: "The question presumes that in order to replicate scientific research, a second researcher has to have access to exactly the same computer program (or 'code') as the initial researcher. This premise is false. The key to replicability is unfettered access to all of the underlying data and methodologies used by the first researcher."

Of course Mike was exactly right, and indeed colleagues at the National Center for Atmospheric Research had done exactly that, writing their own computer code and using the same data that we had, to produce the same results. But Mike wasn't willing to leave this point alone. He threw down a challenge: "It also bears emphasis that my computer program is a private piece of intellectual property. . . . Whether I make available my computer programs is irrelevant to whether our results can be reproduced. And whether I make my computer programs publicly available or not is a decision that is mine alone to make." He went on, "It is a bedrock principle of American law that the government may not take private property 'without [a] public use,' and 'without just compensation.'"

In fact, thanks to diligent efforts by Dave Verardo, our program manager at the National Science Foundation, we knew that we had met all the necessary requirements for transparency and data sharing that were required by that agency. Mike was on solid ground, and he knew it. He quoted directly from a letter Dave had sent to McIntyre and McKitrick, who had dogged our every step in recent months like annoying puppies yapping at our heels. It was clear that Dave had had enough of them too when he wrote to them, stating:

Let me clarify the US NSF's view in this current message. Dr. Mann and his other US colleagues are under no obligation to provide you with any additional data beyond the extensive data sets they have already made available. He is not required to provide you with computer programs, codes, etc. His research is published in the peer-reviewed literature which has passed muster with the editors of those journals and other scientists who have reviewed his manuscripts. You are free to [undertake] your analysis of climate data and he is free to [undertake] his. . . . I would expect that you would respect the views of the US NSF on the issue of data access and intellectual property for US investigators as articulated by me to you in my last message under the advisement of the US NSF's Office of General Counsel.

He might as well have signed it, "Thank you, and good night."

Barton had originally claimed that his request was based on the moral high ground. His letter to Mike stated:

We open this review because this dispute surrounding your studies bears directly on important questions about the federally funded work upon which climate studies rely and the quality and transparency of analyses used to support the IPCC assessment process. With the IPCC currently working to produce a fourth assessment report, addressing questions of quality and transparency in the process and underlying analyses supporting that assessment, both scientific and economic, are of utmost importance if Congress is eventually going to make policy decisions drawing from this work.

The reference to "this dispute" takes us back to the article in the *Wall Street Journal* describing claims by McIntyre and McKitrick that they had "audited" our work, found it to be incorrect, and then "corrected" it. Mike addressed this point head-on:

The various claims of McIntyre and McKitrick—including the ones repeated in your question—have been exhaustively examined by two different groups of climate researchers who have found their objections to be unfounded. . . . Nor is that surprising. *Energy & Environment* is not a peer reviewed scientific journal; it is a journal primarily devoted to policy rather than science that appears to engage in, at most, haphazard review of its articles. And neither McIntyre nor McKitrick is a trained climate scientist. According to the biographical data on their websites, Mr. McIntyre is a mining industry executive with no formal training in any discipline related to climate research and Mr. McKitrick is an economist with no scientific training, hardly credentials that lend force to their academic arguments.

To emphasize his point, Mike continued his take-no-prisoners attack:

The editor of *Energy & Environment*, Ms. Sonja Boehmer-Christiansen, has candidly acknowledged that the publication has a clear editorial bias. In the September 5, 2003 issue of the *Chronicle of*

Higher Education, Ms. Boehmer-Christiansen is quoted as describing the editorial policy of *Energy & Environment* in this way: "I'm following my political agenda—a bit, anyway. . . . But isn't that the right of an editor?" As to "peer review," Ms. Boehmer-Christiansen has acknowledged in an email to Dr. Tim Osborn of the Climatic Research Unit at the University of East Anglia (U.K.), that in her rush to get the McIntyre and McKitrick piece into print for political reasons, *Energy & Environment* dispensed with what scientists consider peer review ("I was rushing you to get this paper out for policy impact reasons, e.g. publication well before COP9"). As Ms. Boehmer-Christiansen added, the "paper was amended until the very last moment. There was a trade off in favour of policy."

Boom! I had a vision of a small Canadian ship disappearing beneath the waves as a massive torpedo hit it dead center. Not that such factual information was likely to mollify the Republican inquisitors; this was after all a political stunt, not an intellectual search for truth. Nevertheless, an objective observer would surely agree that Mike had done a remarkably thorough job of defending himself.

Boehlert's Bombshell

By the time we had figured out how to respond to the House Energy Committee's requests, it was almost mid-July 2005. We had no idea how the committee would respond, given that none of us had acquiesced to their every request. I had little doubt that we would be hauled in to testify, facing assertions that we had fabricated our results and manipulated data to achieve a preconceived goal. Although this was completely absurd, I could see that we would be at a significant disadvantage, with committee members primed to make us look bad. I began to sleep badly and wake up early, tossing and turning as I played out in my mind various scenarios.

I thought about George Galloway, the British MP who had been called to testify before the U.S. Congress about his supposedly illegal dealings with Saddam Hussein. On the face of it, he was being dragged into the media spotlight, where he'd presumably be interrogated and humiliated by Senator Norm Coleman, a Minnesota Republican and a slick ex-prosecutor turned politician. But Galloway completely turned the table on his accusers. "Senator," he told Coleman, "I know standards have slipped over the last few years in Washington, but for a lawyer, you are remarkably cavalier with any idea of justice. I am here today but last week you already found me guilty. You traduced my name around the world without ever having asked me a single question, without ever having contacted me, without ever having written to me or telephoned me, without any attempt to contact me whatsoever. And you call that justice."[6] He blew his inquisitors away and left Washington without any further action being taken. Galloway was my hero, and he was going to be my guide.

I would start off my testimony by alluding to the notorious McCarthy era, when people were dragged before Congress to prove they were not communists. Like the Barton effort to discredit our work, the McCarthy campaign of smear and innuendo was enough to destroy people's careers, regardless of the merits of each case. Senator Joseph McCarthy famously asked each witness, "Are you now or have you ever been a member of the Communist Party?" I decided that I would open my statement by alluding to that sorry episode in congressional history: "I am now and have always been a climate scientist!" When Barton asked me a question, I would respond by saying, "Congressman McCarthy—I mean, Barton . . . " I would refuse to take it lying down; I'd be another Galloway, blazing a path of truth, freedom, and the American way. Then I'd wake up and head downstairs for breakfast.

It's funny how heavy the weight of a government-led inquiry can become. It quickly leads to paranoia. I began to think that my phone was being tapped. This seems silly in retrospect, but there appeared to be no limits to what the government could do. There were daily reports in the newspapers about unauthorized wiretaps, without any legal oversight. Telephone companies had been providing the government access to phone lines, and it seemed quite plausible that Republicans in Congress could slip in a request to keep the three of us under surveillance. This idea became firmly planted in my head when one day, while I was talking to Malcolm Hughes, the phone I was holding suddenly rang in the middle of our conversation. In all the years I've lived in the United States, that had never happened to me before. I also noticed a strange clicking sound when I used the handset. I was sure the phone was being tapped. I called Verizon and demanded that they check to see if my line was being bugged. I was surprised at their casual response: "Certainly, sir. Will Saturday be okay?" When I returned to the living room, Jane looked up. "Oh, by the way, we need a new battery for that phone—it keeps making a weird clicking sound."

I was becoming paranoid and felt precariously alone in dealing with Barton's committee. But the request had not gone unnoticed by others on the Hill. First off the mark was Representative Henry Waxman of California, who happened to be a member of the Committee on Energy and Commerce, albeit a member of the minority party. He issued a letter to Barton in which we were elevated to a wholly undeserved status; we became "three of the world's most respected experts on global warming." Apparently we were climate scientists with "impeccable records." The letter continued heaping praise: "They are widely regarded as among the leading researchers in their respective fields. Collectively, they have published hundreds of articles and dozens of books in the field of earth and atmospheric sciences and received numerous distinctions." Who would have guessed that Henry Waxman had been keeping track of our careers! He expressed dismay that the committee had demanded information "about 'all financial support' they ever received during their long and distinguished careers, 'the source of funding' for every study they ever conducted, 'all data archives' for every published study they ever wrote, and multiple other burdensome and intrusive subjects." He then pointed out the blatantly political nature of Barton's demands, addressing his colleague directly:

Although you have failed to hold a single hearing on the subject of global warming in the eleven years that you have been chairman of the Committee on Energy and Commerce and its Energy and Oversight Subcommittees—and have vociferously opposed all legislative efforts in the Committee to address global warming—your June 23 letters justify your extraordinary demands of these scientists on the grounds that "the Committee must have full and accurate information when considering matters relating to climate change policy. . . . " These letters do not appear to be a serious attempt to understand the science of global warming. Some might interpret them as a transparent effort to bully and harass climate change experts who have reached conclusions with which you disagree.[7]

Yes, indeed, they might well interpret them that way.

Unfortunately, given that Waxman was a Democrat and thus easily written off as a lowlife whiner by the powerful Republicans in charge of Congress, this letter received little attention. To the press it was below the radar, just another skirmish between politicians who were always at loggerheads anyway. But things changed dramatically a couple of weeks later. Our uphill battle with Barton and his allies was transformed overnight by a truly remarkable letter from Congressman Sherwood Boehlert, chairman of the House Committee on Science, who challenged Barton in no uncertain terms: "I am writing to express my strenuous objections to what I see as the misguided and illegitimate investigation that you have launched concerning Dr. Michael Mann, his co-authors and sponsors."[8]

Boehlert's letter was amazing, not just because of its uncompromising and forceful tone, but because, like Barton, Boehlert was a Republican. Not only were two high-ranking Republicans publicly disagreeing, but also one was openly chastising the other. The letter immediately elevated Barton's inquiry to a matter of national visibility. If I had dreamed up the way I would most like to see this matter evolve, I couldn't have imagined anything as good as Boehlert's letter. It was straight out of the George Galloway playbook. "First," wrote Boehlert, "your committee lacks jurisdiction over this matter. Both the National Science Foundation and climate change research come under the purview of the House Committee on Science. This is in no way my central concern about your investigation but I raise it at the outset because it may have legal implications as you proceed."

I couldn't believe what I was reading. Saint George had risen up and was about to slay the dragon. Boehlert continued, evidently now on a roll, "The insensitivity towards the workings of science demonstrated in your investigative letters may reflect a lack of experience in the areas you are investigating." Then came the bombshell that propelled this matter right into the public arena: "My primary concern about your investigation is that its purpose seems to be to intimidate scientists rather than to learn from them, and to substitute Congressional political review for scientific peer review. That would be pernicious." Yes, indeed: pernicious, malevolent, insidious, malicious. But there was more—much more:

> It is certainly appropriate for Congress to try to understand scientific disputes that impinge on public policy. . . . [B]ut you have taken a decidedly different approach—one that breaks with precedent and raises the specter of politicians opening investigations against any scientist who reaches a conclusion that makes the political elite uncomfortable. Rather than bringing Dr. Mann and his antagonists together in a public forum to explain their differences, you have sent an investigative letter to Dr. Mann and his colleagues that raises charges that the scientific community has put to rest, and asked for detailed scientific explanations that your committee undoubtedly lacks the expertise to review. . . . Therefore, one has to conclude that there is no legitimate reason for your investigation. The investigation is not needed to gain access to data. The investigation is not needed to get balanced information on a scientific debate. The investigation is not needed to prompt scientific discussion of an important issue. The only conceivable explanation for the investigation is to attempt to intimidate a prominent scientist and to have Congress put its thumbs on the scales of a scientific debate. This is at best foolhardy; when it comes to scientific debates, Congress is "all thumbs."

Boehlert had really risen to the occasion. He clearly saw the larger implications of what Barton had tried to do. Of course Barton and his committee had no scientific expertise. In spite of their claim that our research was relevant to "the Committee's jurisdiction over energy policy and certain environmental issues," they had never held a hearing on climate change or expressed any interest in the matter until now. Even if we had provided them with all the information they required—which Mike more or less did in his response—they would not have been able to evaluate the information in any meaningful way. Their inability to do so went to the heart of the matter: they had no real interest in the facts; their purpose was intimidation, exactly as their Republican colleague had recognized. But it was one thing to see what was going on and quite another to stand up and state the case so clearly and publicly. Given the fact that on most matters in the Republican-controlled Congress, the majority marched in lockstep and maintained strong discipline in public discourse, Boehlert's letter was heroic. But he had not finished yet:

> The precedent your investigation sets is truly chilling. Are scientists now supposed to look over their shoulders to determine if their conclusions might launch a Congressional inquiry, no matter how legitimate their work? If Congress wants public policy to be informed by scientific research, then it has to allow that research to operate outside the political realm. Your inquiry seeks to erase that line between science and politics. . . . [A]re we going to launch biased investigations each time a difference appears in the literature? I hope you will reconsider the investigation you have launched and allow the scientific community to debate its work as it always has. Seeking scientific truth is too important to be impeded by political expediency.

Boehlert's words should be etched on the portals of every science building in the country, next to a big image of a thumb pushing down on the balance of scientific inquiry:

> **SEEKING SCIENTIFIC TRUTH IS TOO IMPORTANT TO BE IMPEDED BY POLITICAL EXPEDIENCY. WHEN IT COMES TO SCIENTIFIC DEBATES, CONGRESS IS "ALL THUMBS."**
> *CONGRESSMAN SHERWOOD BOEHLERT (R-NY).*

Boehlert's letter propelled our situation onto the front pages of newspapers all over the country. Of course, it was not the issue of the hockey stick that interested editors but the spectacle of one senior Republican publicly chastising another. This was literally unprecedented in the 109th Congress, as the party whips made sure such disagreements never surfaced. The first major article appeared in *U.S. News and World Report* on July 14 under the banner, "Science: Fighting over a Hockey Stick." It correctly identified Barton as "a noted opponent of caps on greenhouse gas emissions."[9]

Other articles quickly followed. The *New York Times* reported the story under the headline "Two G.O.P Lawmakers Spar over Climate Study,"[10] and the *Boston Globe* led with "Testy Exchange Reveals Rift in GOP over Global Warming," noting: "The unusual public tiff between two powerful GOP lawmakers highlights the sharp divide that drives the nation's climate change debate. Barton, along with President Bush and many House Republicans, opposes mandatory curbs on greenhouse gas emissions and questions the science underlying such efforts. Boehlert, who backs limits on carbon dioxide pollution, said he fears such attacks could chill future scientific inquiry."[11]

In an e-mail response to reporters' inquiries, Larry Neal, a spokesman for the Energy and Commerce Committee, hewed to the party line:

> Requests for information are a common exercise of the Energy and Commerce Committee's responsibility to gather knowledge on matters within its jurisdiction. When global warming studies were criticized and results seemed hard to replicate by other researchers, asking why seemed like a modest but necessary step. It still does. . . . Chairman Barton appreciates heated lectures from Representatives Boehlert and Waxman, two men who share a passion for global warming. . . . We regret that our little request for data has given them a chill. Seeking scientific truth is, indeed, too important to be imperiled by politics, and so we'll just continue to ask fair questions of honest people and see what they tell us. That's our job.

No doubt this sarcastic response did nothing to improve the strained relationship between Boehlert and Waxman on one side and Barton on the other.

A few days later a flurry of well-written editorials appeared in newspapers all over the country. The issue had truly caught fire, and Barton was right in the path of the conflagration. The *Denver Post* headlined its editorial "An Attack on Sound Science."[12] I found this to be splendidly ironic, since the term "sound science" had been appropriated by the "global warming denialists" and their principal congressional ally, Senator James Inhofe.

Whenever the denialists alluded to studies that favored their point of view, they would refer to these studies as "sound science." "What we need is more sound science," Inhofe would proclaim, knowing that to the oil lobbyists and his other supporters, the meaning was crystal clear. It was analogous to Fox News appropriating the term "fair and balanced," when much of what the network reported on this issue was about as far from that standard as you could get. Nevertheless, "An Attack on Sound Science" was a good start to a fine editorial. The writer noted that Barton had launched "an attack on science that is so outlandish that even his peers are beseeching him to stop. Nobody would have a plumber evaluate a neurosurgeon, but U.S. Rep. Joe Barton is heeding an 'amateur statistician' who finds fault with the work of professional scientists. Now he's ordered an investigation. It's silly and dangerous—the worry is that the assault against the science of global warming may have a chilling effect in the lab." The editorial continued: "In launching his witch hunt, Barton cites an analysis of the scientists' work by two Canadian critics, neither of whom is a scientist. (One is an economist, the other a mining engineer described in press reports as 'an amateur statistician.') Yet the studies in question were previously analyzed and accepted by true climate scientists with real expertise." This was fun to read, but the last paragraph was the icing on the cake: "Centuries ago, when the Inquisition forced Galileo to renounce his work that proved that the Earth revolves around the sun, legend has it that Galileo muttered: 'Nevertheless, it moves.' Today, Barton might force climate scientists to utter: Nevertheless, it's getting warmer." Wow! In just two short weeks we had gone from obscure climatologists to "three of the world's most respected experts on global warming," and now we were Galileo reincarnated, jousting with the Catholic Church. This was heady stuff.

The next day brought even more intoxicating coverage. Articles and editorials appeared in three major newspapers: the *Washington Post,* the *New York Times,* and the *Philadelphia Inquirer.* The *Washington Post* article, by David Ignatius, was titled "A Bid to Chill Thinking." In it he noted in passing that Joe Barton was "one of the oil lobby's best friends on Capitol Hill." He described the letters Barton had sent to us as having "a peremptory, when-did-you-stop-beating-your-wife tone," adding:

> The political mischief in Barton's probe is that it tries to fuzz the climate debate. . . . [A] growing scientific consensus prompted a "sense of the Senate" resolution last month that "greenhouse gases accumulating in the atmosphere are causing average temperatures to rise at a rate outside the range of natural variability," and that the problem is caused by "human activity." . . . [A] consensus is finally emerging that climate change is a serious global problem and one that is man-made. . . . [Yet] the strategy of Exxon Mobil and other business interests that resist action on global warming has been to maintain the notion that the scientific evidence is shaky.[13]

Ignatius ended by noting that Barton had received $523,099 from energy and natural resource interests for his 2004 congressional race, and $17,500 from Exxon Mobil since 2001. This marked the first time in this saga that anybody had explicitly suggested that his motives might not have been as pure as the driven snow. This point was taken up again by a *New York Times* editorial on the same day: "According to the Center for Responsive Politics, Mr. Barton has also been a leading beneficiary of campaign funds from the oil, gas and utility industries, which have belittled the warming threat and resisted regulatory efforts to control the burning of fossil fuels."[14] In this editorial we had been downgraded to just "reputable scientists." It seemed our meteoric rise to iconic status was to be short-lived. Plans for T-shirts sporting the slogan "Free the Hockey Stick Three" with an image of Galileo on the back were quickly shelved.

A few days after the *New York Times* piece, the *Washington Post* hit the nail on the head with a wonderful editorial titled "Hunting Witches." This investigation was now clearly being seen by the press as nothing more than an industry-driven witch-hunt, with Barton, the industry's puppet in Congress, carrying out the job he had been assigned to do. As the editorial concluded:

> If Mr. Barton wants to discuss the science of climate change, there are many accepted ways to do so. He could ask for a report from the Congressional Research Service or the National Academy of Sciences. He could hold a hearing. He could even read all of the literature himself: There are hundreds of studies in addition to the single one that he has fixated on. But to pretend that he is going to learn something useful by requesting extensive data on 15th-century tree rings is ludicrous; to pretend that it is "normal" to demand decades worth of unrelated financial information from scientists who are not suspected of fraud is outrageous. The only conceivable purpose of these letters is harassment.[15]

As important as the media attention was in exposing the political nature of Barton's request, for the scientific community at large, the issue was seen as far more perilous. If elected officials could use their positions to attack scientific papers and intimidate scientists, where would it end? Today it might be global warming, but tomorrow the focus could be stem cell research, or evolution, or whatever topic might happen to raise the ire of political insiders. Scientists around the world saw the specter of political pressure dictating the direction of science, deciding which results the politicians favored and which were not acceptable to those in power. Of course, such tactics were not new. In the Soviet Union in the 1940s, Trofim Lysenko used the political cover provided by Stalin to determine which scientific activities were acceptable to the state and which would not be allowed. "Lysenkoism" spelled the end of many scientific careers and exposed the dangers of allowing politicians to put their big, uninformed thumbs on the scales of scientific inquiry. Consequently, it wasn't long before scientists and scientific societies across the world began to issue statements decrying the actions by Barton and his congressional allies. On July 7 the European Geosciences Union expressed its concern:

> Rather than exploring the extensive scientific evidence and literature on climate change and the underlying processes through a hearing of experts representing the wide range of national and international scientific institutions engaged in climate research, the Committee Chairman has addressed three individual scientists, questioning their scientific and personal integrity on the basis of a newspaper article. . . . We do not consider personal inquisition of individual scientists as an appropriate way of probing the validity of the general scientific statements in the IPCC Third Assessment Report (TAR), which represents the state-of-the-art of climate science supported by the major science academies around the world and by the vast majority of scientific researchers and investigations as documented by the peer-reviewed scientific literature.[16]

On the same day, the journal *Nature* (which had published our original "hockey stick" paper) ran an editorial noting that "by requesting information on research that does not fit his world view, Barton seems determined to use his political influence to put pressure on the scientific process."[17] I liked this statement but wondered just what Joe Barton's "world view" might be. In light of the information from the Center for Responsive Politics about his career before he entered Congress, it would not be too surprising if Barton's view of the world was simply whatever he could see from the top of a drill rig.

In the United States, similar concerns were being raised. On July 13, 2005, Alan Leshner, CEO of the American Association for the Advancement of Science (AAAS), the largest organization of scientists in the world, wrote to Barton, saying:

> We very much appreciate the Committee's interest in this important field. Your letters, however, in their request for highly detailed

information regarding not only the scientists' recent studies but also their life's work, give the impression of a search for some basis on which to discredit these particular scientists and findings, rather than a search for understanding. With all respect, we question whether this approach is good for the processes by which scientific findings on topics relevant to public policy are generated and used.

He went on: "We think it would be unfortunate if Congress tried to become a participant in the scientific peer-review process itself. More than that, we are concerned that establishing a practice of aggressive Congressional inquiry into the complete professional histories of scientists whose findings may bear on policy in ways that some find unpalatable could have a chilling effect on the willingness of scientists to conduct work on policy-relevant scientific questions."[18]

These statements were quickly followed by a letter to Congressmen Barton and Whitfield (dated July 15, 2005) signed by a group of very distinguished earth scientists, including Mario Molina (Nobel laureate in atmospheric chemistry) and seven other members of the National Academy of Sciences. In this letter the scientists point out:

> The specific findings of Mann et al. constitute only one item among literally thousands of pieces of evidence that have contributed to the present consensus on the serious nature of climate change. While the 2001 report of the Intergovernmental Panel on Climate Change (IPCC) highlighted this work as a useful illustration of our understanding of the impact of fossil fuel–related emissions on climate change, in no way does the report suggest that it is an essential element of that understanding. This understanding has been developed over many years from many diverse lines of inquiry.

They then addressed the pernicious nature of the committee's demand: "Much of the information that you have requested from the scientists involved is unrelated to the stated purpose of your investigation. Requests to provide all working materials related to hundreds of publications stretching back decades can be seen as intimidation—intentional or not—and thereby risks compromising the independence of scientific opinion that is vital to the preeminence of American science as well as to the flow of objective advice to the government."[19] The inclusion of the phrase "intentional or not" seemed more than a little generous to me, but nevertheless I very much appreciated this remarkable letter of support from such an impressive group.

Of course, the writers of these letters weren't saying that they supported the research we had published—the hockey stick graph—nor would I have expected them to do so. The real concern in all of these remarkably polite and restrained statements was that politics had strayed across the line into the domain of independent scientific inquiry. Actually, "strayed across the line" is again far too generous a characterization of what had happened. I would prefer to describe these politicians as deliberately stomping into territory where they don't belong, and in the process abusing their powerful positions in Congress.

Joe Barton was happy to ride this particular horse right back to Texas, where his supporters and financiers were pleased to see him protecting the taxpayers' interests. In a letter to the editor of the *Dallas News* (July 31, 2005), Barton defended his tactics, cowboy-style:

> A few weeks ago, I wrote to three reputable climate scientists and asked them to tell Congress about the facts underlying their theory that human activity is warming the planet. From the spate of political complaints that letter generated, you'd think I had invited them to quit breathing instead of to start bragging. . . . My letter did ask for a lot of facts, and some say it is uncomfortably burdensome. I don't think so. . . . In the end, however, sharing data seems like all indoor work and no heavy lifting. That's the definition of a great job in some places.

I must admit, for all the aggravation that Joe Barton created in my life, I love that phrase "all indoor work and no heavy lifting," and I've adopted it as a mantra in my own research. As I dragged myself up nineteen thousand feet to service our weather station on the ice cap overlooking the crater of Mount Kilimanjaro, I chanted it, step by step. As my colleagues and I struggled to load four hundred–pound Ski-

Doos into the back of a Twin Otter to head off to a remote site in the High Arctic, it was a great teeth-grinder. And as we froze our sorry asses off in a tent in the High Andes of Peru, we hummed the words to distract us from the cold. Oh, yes, we love all that indoor work.

That said, I found the rest of Barton's letter far less amusing:

> It is worth noting that this research is financed by you, if you pay taxes. My own working theory on that relationship goes like this: People who use taxpayers' money need to explain what they're doing with it every now and then. Public money for public science doesn't fall like manna from heaven; it comes from the pockets of working people who earned it. . . . The plant worker in Ennis who contributes some of his paycheck to public science is not a "peer" of the scientist—neither am I—but he is worthy of respect and the occasional explanation.

I don't need lectures from the likes of Joe Barton about respect for hard-working taxpayers—in Ennis or anywhere else. My mother worked long hours in a bakery to give me the benefits of my education, and my father died far too young, before he could even retire from the steel factory where he labored. I understand labor very well; I appreciate hard work, and I respect those who contribute to publicly funded science. What I don't respect are those who masquerade as defenders of the working class, who abuse their position for political purposes, and who try to sully the reputations of those whose ideas they simply don't like.

Scientific Oversight

The challenge to Joe Barton from Congressman Boehlert, together with all the media attention focused on their disagreement, led to an uneasy standoff. The House Committee on Energy and the Environment was not about to let the matter drop, but the controversy had raised the criticisms of our research to such a public level that it was in everybody's interest to have the matter resolved by a fair arbiter.

November 2005. Enter Ralph Cicerone, president of the National Academy of Sciences. This august body is charged with periodically preparing reports on issues of national importance. Although our skinny little hockey stick hardly merited such attention, Representative Boehlert wrote to Cicerone and asked him to convene a panel to evaluate the critical comments we had received. This action bothered me. It was not the role of the National Academy to sit in judgment on a scientific paper; that evaluation should (and inevitably would) take place in the scientific marketplace of ideas—in the scientific literature. If our ideas held water, others would confirm what we had found. If we were wrong, it certainly would not be long before there was a consensus that we had screwed up. Our study had gained prominence mainly by its inclusion in the IPCC report. In effect, it had become a symbol of that report and as such was the target of those who wished to discredit the IPCC's procedures and conclusions. I believe that much of the criticism leveled at our work by politicians was in fact really aimed at discrediting the entire IPCC process. Because we were involved in the report (at different levels as authors or reviewers), our critics tried to make it appear that we had promoted our own work. In reality the report included hundreds of compelling figures, and even if the hockey stick had never been mentioned, the case for human effects on the climate system would have been just as clear. The fact that the hockey stick had been selected by the media as symbolic of that report had nothing to do with us, but it turned our work into a political football. More important, it was just as likely that another iconic image would emerge from the next IPCC report, and so it was important that the Academy of Sciences get ahead of the curve and speak out on the quality and validity of the report's conclusions.

I wrote to Ralph Cicerone expressing all of these concerns. And to my surprise, within an hour of faxing my letter, I received a phone message from him assuring me that this assessment was not going to focus strictly on our papers but would be broadened to address the larger issues surrounding recent climate change.

Accordingly, a panel was established with the task of evaluating the state of research on "surface temperature reconstructions for the last 2000 years." This panel was to be chaired by a distinguished meteorologist, Gerald North, from Texas A&M University, with eleven other well-established panel members broadly representing the community of earth and atmospheric scientists. The intention, at least, was to make the focus wider than the alleged errors in our two papers, but since everybody knew the underlying raison d'être for the panel in the first place, it was hardly a surprise that the hockey stick took center stage.

Numerous scientists were invited to attend the meetings and give presentations. Mike Mann and Malcolm Hughes both went; I declined. I figured it would be a circus, and by all accounts it was. Our critics came out swinging, while Mike, always quick on his feet, bobbed and weaved and counterpunched. Numerous prominent denialists sat in the audience relishing the public bloodletting. I suppose it had to be that way, to give them all their day in court, so to speak, and then close the doors and allow the jury to go about their business in a more sane and thoughtful way. And so they did, producing a 130-page report within a few months—a remarkably fast turnaround for such an eminent and relentlessly busy group.

Their conclusions were—to our eyes—entirely as expected. They found that "the basic conclusion of Mann et al. (1998, 1999) . . . has been supported by an array of evidence that includes both additional large-scale surface temperature reconstructions and pronounced changes in a variety of local proxy indicators . . . which in many cases appear to be unprecedented during at least the last 2000 years." Thus, it was "plausible that the northern hemisphere was warmer during the last few decades of the 20th century than during any comparable period over the preceding millennium." Plausible: "seeming reasonable or probable," from the Latin *plaudere*, "to applaud." Further on, however, the group's report muddled an otherwise startlingly clear message by stating that "less confidence can be placed in the original conclusions by Mann et al. (1999) that the 1990s are likely the warmest decade, and 1998 the warmest year, in at least a millennium."[20] The difference was whether it was "the last few decades" that were exceptionally warm, which they agreed was so, compared to the "last decade," which they did not. I found this bizarre, as nobody questions that the 1990s were at the time the warmest of the past few decades and that 1998 was undoubtedly the warmest year of that decade. If you accept one conclusion, you surely must accept the other. You can't have it both ways.

In any case, the report provided grist for the mills of both hockey stick supporters and our many detractors. No sooner had the report been released than the spin machines began their work. To the denialists, there was no doubt that our conclusions had been resoundingly rejected. But to the rest of the universe, including almost every major newspaper, television station, and scientific publication, we had been vindicated. The BBC announced, "Backing for Hockey Stick." *Nature* reported, "Academy Affirms Hockey Stick." And headlines in the *New York Times* and *Boston Globe* read, "Panel Supports a Controversial Report on Global Warming" and "Report Backs Global Warming Claims," respectively.

As I scanned the media, I was thankful that the negative spin-meisters had not been effective. We had finally been vindicated, and the news media were happy to report an end to the controversy. I was even flown down to New York to appear on CNN's *Lou Dobbs Show,* where the avuncular host congratulated me and Malcolm Hughes on the successful end game.[21] We were out of the penalty box and able to skate freely once again.

But, alas, Congressman Barton was not about to give up. With the unlimited resources of Congress at his disposal, he wanted the last word. It wasn't long before he announced that there would be yet another hearing, but this time he would bring out the big guns. Not content with a National Academy of Sciences report that did not endorse his opinions, he was going to commission a team of statisticians to evaluate carefully the methods used in the Mann, Bradley, and Hughes articles and report back to Congress. He wanted all the egregious errors we had made to be exposed and made part of the congressional record, for posterity. Our statistical sins were to become a small chapter in legislative history, destined to be enshrined in the hallowed archives of the Library of Congress. What an excellent use of the hard-earned tax dollars of all those working men and women back in Ennis, Texas.

Barton asked Edward Wegman of George Mason University and a couple of other statisticians to evaluate whether the criticisms leveled against our work had any legs. The logic seemed to be that since few climatologists are members of the American Statistical Association, they obviously don't know squat about statistics. In fact this is a complete red herring, as the field is replete with scientists who were trained in statistics (some with Ph.D.s from departments of statistics), and there are frequent meetings focused on "statistical climatology." Furthermore, Mike Mann is exceptionally knowledgeable in this area; he was one of the first participants in the "geophysical statistics project" of the National Center for Atmospheric Research in the mid-1990s, and for three years he served on the Committee on Probability and Statistics of the American Meteorological Society. But why spoil a public bloodletting with facts? Bring out a few card-carrying members of the American Statistical Association and let them have at it.

Wegman chose first to examine one important aspect of our research which focused on whether hockey stick–type records could be produced simply by setting 1901–1980 as the baseline for a study of global temperatures over the twentieth century. His committee concluded that because the average for the century was so different from those of previous centuries, a hockey stick shape would always be produced by following our analytical procedures. They argued that what we should have done was to use a longer period as the baseline, perhaps 1600–1900.

Here again we had a problem with critics cherry-picking only part of the procedures we had established in our work and finding fault with that one part. We chose 1901–1980 as the baseline because that was the period when the proxy records overlapped with the instrumental data, an essential step in the calibration process that allowed older data to be converted into temperature estimates. But most important, what we had done was to follow a series of steps, with each step incorporating different data sets, so all of their unique characteristics could be used. The essential point was that you had to continue through *all* of those steps; otherwise useful data would be left out, and indeed you would then produce a meaningless reconstruction of the past temperature record. This is exactly what our two Canadian critics had done. It would be like dismantling a car, then putting it back together again without using some quite important part—like the wheels—and then arguing that the thing doesn't work right. When we broke all the data down into their different components, and then properly reassembled those parts without leaving anything significant out, the hockey stick emerged as clear as day. This has since been demonstrated several times by independent analysts. So Wegman had made a technical point. It just was not relevant to the overall procedures we had followed and hence to the ultimate issue of the hockey stick. Besides, we have since shown that even if you entirely avoid the procedure that Wegman and the Canadians objected to and simply average all the data we used, you get the same hockey stick result. Simply put, the hockey stick is bombproof. No amount of data manipulation will make it go away.

As I read the Wegman report, I must say I was impressed by how well this statistician had grasped the intricacies of paleoclimatology and, in particular, high-resolution studies of tree rings, ice cores, and corals. His section on the problems of using tree rings, and of the important points that one must take into account, struck me as quite brilliant—lucid and clear. It was only later that I realized that large sections of his report had been lifted verbatim from my own 1999 book on the subject, *Paleoclimatology*. I don't think the word "irony" does justice to the fact that a person commissioned by Congress to investigate the wrongdoings my colleagues and I had supposedly committed had the nerve to reproduce entire paragraphs from a book written by one of the people under investigation, without even citing the source.[22] In academic life, plagiarism—using what others have written and passing it off as your own—is the ultimate sin. The American Association of University Professors has published a clear statement on the matter: "Plagiarism is . . . the antithesis of the honest labor that characterizes true scholarship and without which mutual trust and respect among scholars is impossible." At Wegman's university, the student honor code clearly warns that plagiarism may be grounds for expulsion, noting, "Plagiarism is the equivalent of intellectual robbery and cannot be tolerated in an academic setting." Perhaps it is asking too much for the United States Congress to meet the standards that universities set for their undergraduates, but I did expect a bit more honesty from the Bernard J. Dunn Professor of Information Technology and Applied Statistics.

Wegman, however, went a step further. In a bizarre departure from his brief to evaluate our statistical procedures in climate reconstruction, he decided to undertake a "social network analysis" of papers published by Mike Mann and other colleagues. He discovered that Mike (being an exceptionally productive scientist) had published papers with a lot of different coauthors. So had I, and so had Malcolm Hughes. After all, we were considerably older than Mike and so had had longer to accumulate collaborators. Sometimes Mike's coauthors were the same people Malcolm and I had also worked with earlier in our careers. This is hardly a surprise. When you undertake a study that involves different kinds of data, you might well team up with a particular person so as to benefit from the detailed knowledge that person has, and give credit for that person's contribution to the work. This is normal in all aspects of scientific inquiry.

Wegman's analysis went far beyond this obvious fact. His report implied that if I had published a paper with somebody, then later on, when asked to review a paper or proposal by that person, I would be incapable of making an independent assessment of his or her research. In effect, Wegman suggested, we were all members of a cozy club, a group of sycophantic, self-serving good old boys and gals who blindly endorsed whatever landed on our desks, as long as it came from a former coauthor. As Mike Mann nicely noted, in a response to Wegman's report, "Wegman claims that there is, in essence, an almost sinister conspiracy of like-minded climate scientists who act as a cartel to control the published literature in climate studies."[23]

The reality is somewhat different. Science is a very competitive business, and scientists are trained to be skeptical and to critique the work of others. I am quite sure that Mike wonders how some of the things I have written passed peer review (a belief I base on the fact that I feel the same way about some things he has written). So it goes. No hard feelings (we hope); we just move on. In any case, in the marketplace of results (that is, in the peer-reviewed literature), when papers of marginal value, or even poor science, do occasionally get through the review process, more often than not this is because the reviewer or journal editor did not give the paper enough thought or time, not because it came from a buddy. Peer review is a necessary but sometimes not always sufficient criterion for good science to appear in print. And when such marginal papers are published, it's not too long before another paper comes along critiquing it and correcting the record. In this way science stumbles forward, slowly illuminating the path toward some universal truth (again, we hope).

As for research proposals, the various science agencies that evaluate them go to extraordinary lengths to avoid any possibility of bias in a review. After all, these involve real money! Anyone submitting a proposal to, say, the National Science Foundation must list everybody with whom he or she has collaborated in the last few years, as well as all former graduate students, postdoctoral advisees, former advisers, and so on. This list then guides the program managers as they select referees to evaluate the proposal. In fact, this is not always the ideal procedure; in some cases the program managers would like to get an opinion from the person who has the most expertise on a subject, but that person turns out to have been an adviser of the young scientist who is now trying to obtain independent funding for his or her own research. On the whole, however, the system seems to work well, and good science is rarely turned down, either for funding or for publication. I've had my share of disappointments on both fronts, but after more than thirty-five years in this business, I can say that it is remarkably fair. Or perhaps I should say that, like democracy, it's the worst possible system—except for all the other options.

The irony of Wegman's criticisms is obviously lost on politicians. In the ultimate club—the U.S. Senate—members routinely pillory their opponents, denouncing their positions, questioning their moral fiber and patriotic fervor. But occasionally, when it suits them, they will sign on to (or perhaps even cosponsor) the same legislation. They may even go out for dinner, or play a round of golf with their "honorable friends." It is truly remarkable to suggest that only politicians are able to partition their opinions in such a clear and unbiased way, whereas the rest of us are incapable of objective criticism, caught up in a "social network" of people who simply flatter us and approve our every word.

Reading the transcript of the hearing convened by Congressman Barton at which Wegman testified about his report (July 19, 2006), I was struck by one particular exchange. As Wegman tried to explain some of his findings, Congressman Jay Inslee, a Democrat from the state of Washington, interrupted him:

> I want . . . to make sure you understand the reality of the situation. I am giving to you all the sincerity that I can give to you. But the reason you are here is not why you think you are here, OK? The reason you are here is to try and win a debate with some industries in this country who are afraid to look forward to a new energy future for this nation, and the reason you are here is to try and create doubt about whether this country should move forward with a new technological clean energy future, or whether we should remain addicted to fossil fuels. That is the reason why you are here.[24]

And of course he was exactly right. The goal of Congressman Barton, Senator Inhofe, and others like them is to ensure that legislation to control greenhouse gases is never passed by the U.S. Congress. Their strategy, like that of the tobacco industry in the past, is to sow the seeds of doubt about climate science, and if that means destroying the reputations of those who carry out the science, so be it.

The moral of this sorry tale is that politics and science must not be allowed to mix. It is far too easy for political considerations to overwhelm the issues at stake, and then to treat scientists like political opponents. Politicians are elected to make policy, and to do that well, they need access to good, unbiased scientific information. It is extremely rare that scientists can be absolutely certain about an issue; there are almost always questions that require further research. As global warming deniers are fond of saying, science is never settled. That does not mean, however, that scientific issues are always balanced on such a knife-edge of uncertainty that decisions can't (or shouldn't) be made. Today there is an overwhelming scientific consensus that further increases in greenhouse gases will lead to disruptive changes in the global climate system. In IPCC terminology, this outcome is "very likely."[25] Some argue that science should not be based on consensus; otherwise we might still be following the edicts of the Catholic Church that the sun revolves around the earth. Science needs critics (like Galileo) to test and question what we think we know. Scientific skepticism is healthy, but a stubborn refusal to accept facts is not. Today we are faced with something even worse—a deliberate rejection, based on political ideology, of even the most obvious and unequivocal observations that the earth is getting warmer. Global warming is thus "just a hoax." Apparently glaciers have been fooled into receding, sea level has been deceived into rising, plants and animals have been tricked into migrating ever poleward. I don't think so. As Senator Daniel Patrick Moynihan famously said, "Everyone is entitled to his own opinion, but not his own facts."

It is entirely possible to accept that there will always be uncertainty but nevertheless still take very seriously important scientific issues that might affect everyone on earth. Politicians operate in an environment of uncertainty all the time; economic decisions that affect us all are made every day on the basis of economic models and projections of the future. Why, then, should we ignore the advice of those who have studied the matter of global warming carefully, who have questioned and tested and challenged one another, yet who nevertheless return again and again to the same stark conclusion. According to the 2007 IPCC Assessment Report:

Warming of the climate system is unequivocal. . . . most of the observed increase in global average temperature since the mid-20th century is *very likely* due to the observed increase in anthropogenic greenhouse gas concentrations. . . . [C]ontinued greenhouse gas emissions at or above current rates would cause further warming and induce many changes in the global climate system during the 21st century that would *very likely* be larger than those observed during the 20th century."[26]

CHAPTER THREE

4

The IPCC and the Nobel Prize

Looking back on the whole experience with Congressman Barton and his henchmen, and Senator Inhofe and his bizarre obsession with the hockey stick graph, I suspect that their main concern was not so much the story the graph told but rather the IPCC report. Because the graph was highlighted in the summary report of IPCC's Third Assessment Report, and was then adopted by the media as an eye-catching image to represent the report's findings, the idea of killing off the graph became synonymous with discrediting the entire IPCC. In addition, Barton and others knew full well that the next IPCC report—the Fourth Assessment—was already under way (it appeared in 2007), and it was unlikely to come to conclusions very different from those in the Third Report. In discrediting the hockey stick, and by extension the IPCC, their aim was to blunt whatever publicity might accrue to the next report.

This was not the first time that entrenched interests had waged war on the IPCC, and the tactics they employed in the case of the hockey stick (and its authors) were straight out of the playbook that was used earlier against another scientist, Ben Santer. Disliking the message of the IPCC, they sought to destroy the reputation of one of the messengers. As in our case, a despicable campaign of misinformation, misrepresentation, and character assassination was employed to discredit Ben's honesty and question his scientific integrity. In Ben's case, the instigators were individuals aligned with the Global Climate Coalition, an energy industry–sponsored group that included Exxon-Mobil, Texaco, Royal Dutch Shell, Ford, GM, and DaimlerChrysler. Their goal was to oppose legislation aimed at controlling greenhouse gases. In our case, the players were elected officials, well funded by some of the same industries, who used their powerful congressional positions to disseminate misinformation and intimidate individuals. But behind all the drama, the work of the IPCC proceeded. It is worth taking a minute to consider what that work was all about.

Much has been written about the IPCC, but it remains a rather misunderstood organization. It was established in 1988 under the auspices of the United Nations, with the goal of assessing "the scientific, technical and socioeconomic information relevant for the understanding of the risk of human-induced climate change."[1] To accomplish that task, a small army of scientists are periodically asked to review all the published literature up to a particular date, and to summarize the "state of knowledge" at that time. In that sense, "the panel" does not undertake any original research; its members' job is to read, review, and assess work that has already been done, using just the published literature.

The panel consists of three "working groups": Group I focuses on the basic science of climate change ("the scientific aspects of the climate system and climate change"); Group II examines climate change impacts, adaptation, and vulnerability ("the vulnerability of socioeconomic and natural systems to climate change, negative and positive consequences of climate change and options for adapting"); and Group III addresses issues related to the mitigation of climate change ("options for limiting greenhouse gas emissions and otherwise reducing the impacts of climate change").[2] Another group is charged with estimating how greenhouse gases might increase in the future under a range of possible scenarios. This group produced the "Special Report on Emission Scenarios" (SRES), and those projected changes are an important yardstick for climate modelers as they attempt to predict how climate may change in the future under the different scenarios envisioned by that group.

Given the politics of climate change, the IPCC has been scrupulously organized so that its procedures are about as clean and uncontaminated as an operating room. Because it is an "intergovernmental" organization, the entire process begins with a meeting of delegates representing 194 different countries. At this meeting the scientific leader of the forthcoming IPCC assessment is elected, along with three vice chairs, and the chairs and vice chairs of the working groups and a Task Force on National Greenhouse Gas Inventories. This adds up to thirty people, but they must be equally distributed among six geographical regions (more or less representing Africa; Asia and the Middle East; South America; the South Pacific Islands and Australasia; North and Central America and the Caribbean; and Europe). In fact, each working group must have two leaders, one from a "developed" and the other from a "developing" country, to provide balance.

One could argue that, given the level of climate research in Europe and North America compared to, say, in Africa, this seems an odd way to start. But this is an intergovernmental panel, so that's how it operates. And in fact it is important to ensure that all countries are involved in the assessment. Although it makes for some strange bedfellows, so to speak, the process seems to work quite well. It has certainly made it possible for scientists in the less developed countries to have a voice at the table, and even though they may not have published a large number of the primary articles on climate change, an effort has been made to involve as broad a set of people as possible in this "research assessment" exercise.

The first IPCC report came out in 1990. Working Group I was led by three scientists from the UK Meteorological Office, but the eleven chapters of the report represented the joint effort of several hundred scientists from twenty-five different countries. Of course, trying to get a crowd that large to write a document would be a hopeless task. So the process begins with numerous reports being submitted by "contributing authors," each summarizing the latest research on a particular aspect of the science with which they are most familiar. These reports are then synthesized by a smaller group of "lead authors" (typically fewer than five people). Each draft chapter is then circulated among the contributing authors for comments—to see if it properly summarizes the information, or if it has under- (or over-)emphasized an important point. Once everyone is satisfied that the chapter is ready for a wider audience, it is sent to a set of expert reviewers, and their comments are then considered and appropriately digested in a revised version of the chapter. Finally, the revised drafts are posted on the Web for further review and comment by governments and other "interested parties." At each step the lead authors may revise the text accordingly. If they decide to make no change, they explain to the person who made the suggestion why no action was taken. In this way the latest science is distilled from the published literature, and misstatements and misunderstandings are caught; at least, that is the hope. In the Third Assessment Report, this whole process involved 122 lead authors, 515 contributing authors, 21 review editors, and 337 expert reviewers.

Wacky ideas that may have made it through the publication process into different journals will probably not survive this level of scrutiny. It is left to the best judgment of those involved—particularly the lead authors—to decide which ideas and new data sets have added to our knowledge base since the preceding IPCC report. In this sense the notion that the IPCC represents a "consensus" of all those involved is not quite on target, though all of the lead authors agree to stand by their final document. I think of the process as rather like the operation of a large restaurant. Vendors (contributing authors) may be asked to stop by with a selection of their latest produce. Each will extol the virtues of all the fresh food recently culled from the garden. The chefs (lead authors) will inspect it all but may decide that those tomatoes aren't quite ripe yet, or the cabbages are a bit past their prime. Only the really good stuff will be used to create the day's menu. Ultimately the choices are made by the chefs, but all involved have a chance to argue their case and try to get their basket of produce adopted.

As the grand feast is prepared, disagreements may persist further down the food chain, but the lead authors nevertheless aim to present the best of current climate science literature. Inevitably there will be some work that simply does not rise to the level of significance in their eyes. But really important material tends to persist and be worked on more, so key issues will very likely show up in the next report if they "have legs." The process is not perfect, but it is remarkably open, and those involved (in my experience) are eager to "get it right." Furthermore, each chapter is then released in draft form so that others may read, critique, and comment on the text. All these reviews are published on the Web, and every comment is responded to. This huge effort is designed to make the entire process as open and accessible as it could possibly be, and to ensure that nothing has been overlooked or given inadequate consideration.

The tricky part is what happens after the scientific assessments have been completed. The assessments are often voluminous tomes which no policymaker will ever read. A "Summary for Policymakers" is always prepared for this reason. Clearly, when you try to boil down several hundred pages into a few choice words and figures, a lot has to be omitted. Furthermore, these summaries must be approved by the governments of all the countries involved, so there is a lot of arm-wrestling that goes on, with every sentence carefully weighed, literally word by word, to ensure that it does not stray too far from what is scientifically justified. It is in these discussions that the scientific chapters are approved for public release by all the government representatives, and final changes in the "Summary for Policymakers" are approved. This brings us back to the story of Ben Santer. Poor Ben had the misfortune of being the lead author on chapter 8 of the IPCC's Second Assessment Report. The product of Working Group I, this chapter focused on the "detection of climate change and attribution of causes." These terms have very specific meanings, as the report noted:

> Any human-induced effect on climate will be superimposed on the background "noise" of natural climate variability, which results both from internal fluctuations and from external causes such as solar variability or volcanic eruptions. Detection and attribution studies attempt to distinguish between anthropogenic and natural influences. "Detection of change" is the process of demonstrating that an observed change in climate is highly unusual in a statistical sense, but does not provide a reason for the change. "Attribution" is the process of establishing cause and effect relations, including the testing of competing hypotheses.[3]

Ben was given the task of figuring out what happened and whodunit. Fortunately, the effects of different factors leave a distinct "fingerprint" on the climate system. For example, the energy emitted from the sun warms the surface and the atmosphere at all levels. But the fingerprint of greenhouse gases is quite different. They warm the surface and lower atmosphere but cool the stratosphere (the uppermost atmosphere). Other factors—such as volcanic aerosols—cool the surface but warm the upper atmosphere; and so on. There are also seasonal differences and changes in the geographic patterns of climate change attributable to all of the major climate-influencing factors.

Researchers, then, must look for the specific fingerprint of greenhouse gases in the climate data to see if that pattern is becoming clearer over time (see chapter 3). That was the focus of the "detection and attribution" chapter: Was it possible to detect the fingerprint, or signature, of greenhouse gas influences on the world's climate? Ben and the other lead authors were cautious, pointing out the difficulties of doing this: "Our ability to quantify the human influence on global climate is currently limited because the expected signal is still emerging from the noise of natural variability, and because there are uncertainties in key factors. These include the magnitude and patterns of long-term natural variability and the time-evolving pattern of forcing by, and response to, changes in concentrations of greenhouse gases and aerosols, and land surface changes." Notwithstanding those limitations, it looked as if a picture was emerging, so they concluded with this important statement: "Nevertheless, the balance of evidence suggests that there is a discernible human influence on global climate."[4]

The combination of cautionary language followed by a tentative conclusion was very similar to the approach used in the "hockey stick" paper, which we published a few years after Ben's chapter came out in 1996. As I have already noted, our paper was titled "Northern Hemisphere Temperatures during the Past Millennium: Inferences, Uncertainties and Limitations," so nobody should have missed the red flags. In both cases, though, all of the caveats and words of warning went out the window; the media just focused on the final phrases.

When Ben presented the results of his chapter to government representatives at an IPCC plenary meeting in Madrid, which was assembled to approve the whole Second Assessment Report, there was criticism that parts of his chapter needed further clarification to address issues raised by some reviewers. Some of those in attendance also felt that a "concluding summary" was not needed since the chapter already had an "executive summary," and no other chapter had such a concluding section. They told him just to clean it up a bit, which Ben dutifully did. He was in fact *required* to do this under the IPCC rules.

Six months later, on June 12, 1996, an op-ed article appeared in that famous scientific resource the *Wall Street Journal,* titled "A Major Deception on Global Warming." The author was Frederick Seitz, a well-known physicist with stellar credentials. Seitz had been president of Rockefeller University and president of the National Academy of Sciences, and had received the National Medal of Science. He was no slouch, scientifically, though it is important to note that he was also chairman of the George C. Marshall Institute, a right-wing "think tank" focused on scientific issues and public policy. The most vocal opponents of legislation to control greenhouse gases are either on the board of directors of the Marshall Institute or in some way affiliated with it.

Seitz's orientation on the global warming issue should thus have been clear; nevertheless, his article received immediate media attention. In it he argued that chapter 8 of the IPCC's Second Assessment Report had been altered "after scientists charged with examining this question [of the human influence on climate] had accepted the supposedly final text." And he laid the responsibility for those changes firmly at the foot of the lead author, Ben Santer. Seitz was outraged: "In my more than 60 years as a member of the American scientific community . . . I have never witnessed a more disturbing corruption of the peer-review process than the events that led to this IPCC report." He then dropped his cover a bit, arguing that if the IPCC reports "lead to carbon taxes and restraints on economic growth, they will have a major and almost certainly destructive impact on the economies of the world." Clearly Ben had a lot to answer for: his edits would essentially cause the world economy to collapse.

Reaction was swift. All the usual lapdog bloggers and petulant whiners emerged to add to the furor and express their disgust. Fred Singer, who never misses an opportunity to put the boot in, followed Seitz with a letter to the *Wall Street Journal* (July 11, 1996) claiming that the IPCC was on a "crusade to provide scientific cover for political action."

Understandably, Ben was shaken by these charges. He responded in a letter to the editor (June 26, 1996), co-signed by forty other scientists from eight countries, all lead authors or contributing authors in Working Group I. They noted that all the procedural rules of the IPCC had been followed, and in fact the changes had been made in direct response to written comments offered prior to and during the Madrid meeting. "There has been no dishonesty, no corruption of the peer-review process and no bias—political, environmental or otherwise."

The letter went on to point out that Seitz is not a climatologist, was not present in Madrid, had never participated in the IPCC process, had never seen any of the reviewers' comments, and, by the way, had never bothered to contact anybody involved to see if his damning conclusions were correct. But again, that's typical of the way the obstructionists operate: never spoil a good public bloodletting with facts. This letter was followed by another, signed by Bert Bolin, chairman of the IPCC, and Sir John Houghton and Luiz Gylvan Meira Filho, co-chairmen of Working Group I, who noted that Seitz's allegations "have no basis in fact." They said that they were "completely satisfied that the changes incorporated in the revised version were made with the sole purpose of producing the best possible and most clearly explained assessment of the science and were not in any way motivated by any political or other considerations." Seitz's "attack on Dr. Santer and the other scientists involved is therefore completely unfounded."[5]

Like the hockey stick story, the attacks on Ben Santer nevertheless continued unabated for many months, distracting him from his important research. The campaign against him was fueled by right-wing bloggers and others who couldn't have cared less about the specific issues involved but simply saw this as a way of generating some heat on the issue of global warming. It always helps with such a constituency if you can claim in some way that the United Nations—that shadowy organization bent on world domination—has had a role in corrupting science for its own political objectives.

The happy postscript to this story is that Ben Santer has continued his fine work as a research scientist at Lawrence Livermore National Lab and has made key contributions to the science of detection and attribution of global warming, including much more work for the IPCC. He also received a MacArthur Foundation Award (dubbed "genius awards" by the media), which clearly sent the message that he is a first-class scientist as far as his peers are concerned. But most significantly, the tentative conclusions of the 1995 IPCC assessment that "the balance of evidence suggests that there is a discernible human influence on global climate" has stood the test of time.

In later assessment reports the wording has gradually become more definitive. The 2001 report declared, "There is new and stronger evidence that most of the warming observed over the last 50 years is attributable to human activities."[6] In the 2007 report the corresponding statement read: "Warming of the climate system is unequivocal. . . . Most of the observed increase in global average temperatures since the mid-20th century is very likely due to the observed increase in anthropogenic GHG concentrations."[7] In the precise world of the IPCC, words and phrases take on very specific meanings. "Likely" means that a statement has a 60 percent chance of being true, whereas if the report says that something is "very likely," the probability is greater than 90 percent. In other words, the 2007 IPCC assessment states that, although there is a slim chance that the observed warming is due to something else, that chance is getting smaller all the time, and we are now pretty sure that humans did it.

I began this chapter by linking the IPCC to the hockey stick. So what, exactly, is the connection? In the 2001 Third Assessment Report, Working Group I devoted a very small section to the paleoclimatic evidence for global warming. There were fourteen chapters and eight appendices in that report. Chapter 2, "Observed Climate Variability and Change," contained thirty-four subsections. The hockey stick was described in one of those small subsections. In a report of over 880 pages, it occupied less than one page. There were more than two hundred figures in the book; the hockey stick figure was only one of them. It is quite obvious, then, that our graph was not the basis for the IPCC's conclusions. If the hockey stick had never been included, the physical arguments for human-induced global warming would still have been compelling. The idea that in some way the entire edifice of the IPCC assessment rested on this one figure is simply laughable. Yet that was the view that demagogues like Senator Inhofe attempted to project. Ironically, in doing so they demonstrated their complete and utter ignorance of the scientific basis for concern about global warming.

The only important thing that the hockey stick provided was a longer-term perspective on the warming that was observed all over the world in the twentieth century. It allowed scientists to assess whether the recent changes are just due to a natural cycle, or if there had been warmer periods in the past. It also helped scientists to see if there was any evidence that the rate of change of warming in the twentieth century was unprecedented over the course of the past one thousand years. These were the very questions we had set out to answer in our research, and so it was natural for our results to be included in the assessment. As the report noted: "Globally it is very likely that the 1990s was the warmest decade, and 1998 the warmest year, in the instrumental record (1861–2000). The increase in surface temperature over the 20th century for the Northern Hemisphere is likely to have been greater than that for any other century in the last thousand years."[8]

Much has been made of the fact that Mike Mann was a lead author of this chapter, with the implication that he somehow forced this figure into the limelight. Mike is a persuasive character, but I think that even if he had not been involved in this chapter at all, it would almost certainly have mentioned the research we had done because it was so relevant to the issues at hand. And the hockey stick graph, with all of its uncertainties appropriately stated, is simply a visualization of the points made in the chapter. It would have been an obvious choice to include it, whoever wrote the chapter. In any case, given the rigorous peer review that all chapters underwent, if there had been any serious reservations about including it, they would have been given voice and discussed by all the chapter's authors. The graph would not have survived such a mutiny.

Survive the hockey stick did, and so compelling was the point it made that those involved in writing the overall IPCC "Summary for Policymakers" selected it as a useful image to help readers understand the nature of recent changes in temperature. From there it leaped to the front pages of newspapers and magazines as an iconic symbol of the IPCC's conclusions. A picture is worth a thousand words, or in this case many thousands. To its detractors, the hockey stick came to represent all that the IPCC stood for, even though everyone involved in the process knew that there was a great deal more in the IPCC reports than a graph shaped like a piece of sports equipment.

The IPCC was an irritation that those who were bent on continuing the unregulated consumption of fossil fuels were determined to destroy. Then in 2007 they suffered an unexpected setback: the IPCC was awarded the Nobel Peace Prize, shared jointly with Al Gore. The prize was presented, as the citation read, "for their efforts to build up and disseminate greater knowledge about man-made climate change, and to lay the foundations for the measures that are needed to counteract such change."[9]

With this award in the IPCC's hands, the right-wing fringe exploded into paroxysms of rage. That paragon of well-balanced humility Rush Limbaugh dismissed the honor, noting (as did many other right-wing commentators) that Gore thus joined the ranks of other lowlife recipients such as Yasser Arafat and Jimmy Carter. The notion that global climate change might be considered a threat to society and to peaceful coexistence on the planet was simply not on their radar screens.

The U.S. military, however, had certainly recognized this prospect. In a 2007 report of the Military Advisory Board, "National Security and the Threat of Climate Change," a panel of distinguished military officers came to these stark conclusions: "Projected climate change poses a serious threat to America's national security. . . . Climate change acts as a threat multiplier for instability in some of the most volatile regions of the world. Projected climate change will seriously exacerbate already marginal living standards in many Asian, African, and Middle Eastern nations, causing widespread political instability and the likelihood of failed states." In fact, the panel noted that "projected climate change will add to tensions even in stable regions of the world."[10]

These observations were not the views of left-wing environmental wackos whom Limbaugh and his cronies like to portray as leading the nation astray. These views came from retired flag and general officers from all four services, including service chiefs and some who had served as regional combatant commanders (four-star officers who command all U.S. forces in a given region of the world). As it turns out, the Nobel board was right to acknowledge that those who sought to bring the issues of climate change and global security to the public's attention, through both low-profile but relentless scientific research (the IPCC) and high-profile advocacy (Al Gore), were equally worthy of recognition. The award served as a fitting rebuke of all the shenanigans that had been directed at Ben Santer, the hockey stick team, and the IPCC. Justice prevailed.

5
Global Warming
A Primer

"It's time for us to start talking about 'climate change' instead of global warming. 'Climate change' is less frightening than 'global warming'; 'global warming' has catastrophic connotations attached to it; 'climate change' suggests a more controllable and less emotional challenge." This was the advice given by the Luntz Research Company to Republican candidates running for election in 2003.[1]

Words matter. Politicians who are concerned about the fact that Americans are becoming more favorably inclined toward the regulation of greenhouse gas emissions avoid using the term "global warming." It's all just "climate change." Everyone knows that climate changes naturally, so if the public can be convinced that rising temperatures are just part of the normal pattern, they won't be so worried. If they aren't worried, controlling greenhouse gases won't be a priority, and if it's not a priority, limits on energy use won't be considered.

To me, "global warming" is a term that simply describes the kind of change we've seen over the last century, and especially over the last thirty or forty years. It is part of climate change.

In the past there have been natural changes in climate that were completely unrelated to human influences. But increasingly over the last century, the composition of the atmosphere has changed as a result of human activities, and this has directly affected global temperatures. Initially atmospheric pollution was just a local problem (such as the smoke-related fogs that used to darken the skies of London and other large cities in the nineteenth century), but as time passed, such local effects expanded to become a much more significant global-scale problem. The real problem was not the very visible "smogs" that blanketed so many cities (mainly from the burning of dirty coal); it was the completely invisible (colorless and odorless) byproduct of that combustion, carbon dioxide. I often wonder if we would be struggling with the carbon dioxide problem today if it were a stinking purple gas that would more readily grab people's attention. Alas, its presence is just not very obvious as we go about our daily lives.

Carbon dioxide is a "greenhouse gas" which plays a critical role in maintaining the temperature of the earth at levels we humans (and other forms of life) have come to expect. More than a century ago, the Swedish scientist Svante Arrhenius predicted that if greenhouse gases increased in the atmosphere, global temperatures would rise (by 5–6°C if CO_2 levels doubled). Indeed we have seen a steady increase in temperatures around the world over the last one hundred years as the levels of carbon dioxide and other greenhouse gases have increased. The term "global warming" refers to this increase in temperature, but it also implies that the change is related to human activities that have led to higher levels of greenhouse gases in the atmosphere. Some part of the observed increase in temperature may be entirely natural, unrelated to human activity. Many politicians who do not support legislation to reduce fossil fuel use (generally those supported by energy-related industries) emphasize this natural explanation for temperature increases and avoid using the term "global warming" altogether. The words carry too much baggage and have huge political implications. They think that terms like "climate change" are fine but "global warming" isn't, even though they are part and parcel of the same problem.

Although there are still diehards out there who continue to insist that global temperatures simply are *not* rising (for whatever reason), the facts overwhelmingly demonstrate that they are. I've been involved in studies of climate change for a long time, and while there are lots of topics that could justifiably be classified as controversial, the observed increase in global temperatures is not one of them. Let's be more specific about this: When we say "global," what are we referring to? How much warming has there been, and over what period of time?

The term "global warming" always refers to air temperature. But as we'll see, there are lots of other changes that are related to air temperature, and these confirm that temperatures have indeed risen over the last one hundred years or more. But the air (or atmosphere) extends many miles above the earth's surface. So what exactly is warming?

Although there is clear evidence that the entire lower half of the atmosphere has been getting warmer, when climatologists talk about the record of global warming, they are generally referring to air temperatures near the surface. These are measured in well-ventilated boxes about five feet above the ground; the box provides shade for the thermometers being used to record temperature so as to avoid the effects of direct solar heating. Such measurements have been made at sites all over the world for the last three hundred years or so. In the early days of temperature measurements, most of the observations were made in Europe; it took a long time before similar measurements were made in the Americas, Africa, Asia, Australia, and in polar regions. Consequently, if you want a *global* picture of how temperatures have changed, based on thermometer readings, it's really only possible to do this for the past 150 years. Before that, the distribution of temperature observing stations was just too uneven, leaving large regions unrepresented.

Of course, "planet earth" is to a large extent (70 percent, actually) "planet water," so it is also necessary to have a network of temperature measurements from the ocean regions of the world. To some extent, records from remote islands (such as the Azores, Fiji, and Hawaii) provide data that are representative of the ocean areas surrounding them, but there have also been many measurements made on board merchant and naval ships, which provide records of sea surface temperatures over vast swaths of the world's oceans. No doubt some resentful sailor was assigned the task—to throw a bucket over the side, bring up a water sample, and stick a thermometer in the bucket. He probably could not have imagined that, centuries later, climate historians would pore over each ship's log to extract these crucial bits of data, transferring them like tiny gems to create a bright mosaic of information about the surface temperature of the oceans. Collectively such data have provided a comprehensive picture of temperatures across the seas and the changes that have taken place over many decades.

As you might expect, when you begin to assemble all these records, you run into lots of problems. I immersed myself in these data when I first began studying climatology. My Ph.D. involved a study of weather records from military forts in the western United States which were built in the nineteenth century to pacify the "Wild West." At that time there was a belief that diseases afflicting the troops were related to weather conditions, and as a result, each surgeon general in those far-flung military outposts was required to keep detailed weather records. These reports—now safely filed away in various archives—provide the earliest temperature records for much of the western United States, going back to the 1830s and 1840s in some cases. Very detailed measurements were made every hour at these locations, throughout the day and night, interrupted only when the forts were under siege by angry locals (who presumably did not appreciate the importance of maintaining a continuous set of records). This expansion of weather observations into new territory was also happening in other parts of the world, so that by the mid- to late nineteenth century, there was essentially a global network of data being recorded on a daily basis.

Unfortunately there was no commonly accepted protocol or standard procedure for taking the temperature measurements. Thus in Spain (and in all Spanish colonies), the temperature might have been observed at dawn, noon (before a siesta), and sunset, whereas in Great Britain (and all of its colonies), the measurements might have been made at 9 am (after breakfast), 2 pm (before tea), and 7 pm (after a gin and tonic), while in other countries they might be taken at any hour in between. Obviously this made comparisons between different regions rather problematical. In fact the situation was quite chaotic until an agreement was reached around 1910 to adopt a common standard: daily temperatures would be recorded as the average of the highest and lowest value in each twenty-four-hour period. Today the daily mean (or average) temperature is exactly that for each location, and the monthly mean is the average of all the daily mean values. Similarly, the "global mean" is the average of all the individual values from stations around the world.

Some areas contain far more measurement stations than others (for example, there are not many in the Sahara versus a lot in western Europe). To achieve a fair balance, then, the global average also involves careful selection of data to avoid overweighting one area in comparison to another. After all, we want to get a picture of what is happening across the globe *as a whole*. This is usually done by placing an equally spaced grid over the globe and figuring out what the representative value is for each corner of each grid box, based on values from all the nearest stations. That way, you get an evenly spaced network of data for each month that can be compared with those for subsequent months without any significant bias. Finally, because the number of station records has changed over time, and some are at higher elevations (and thus colder) than others, simply averaging all the available values for each month or year would not be very meaningful. What we want to know is: How have temperatures changed over time, unbiased by changes in the observational network? Each record is thus "referenced" to a common period (ordinarily the average for 1961–1990), and the data are expressed as differences (or anomalies) during that period at each station, month by month and year by year. When all these values are averaged together, the result is a graph that shows variations over time in relation to the average for the common baseline at all locations.

To extend the records back in time requires painstaking adjustments of each station's record to take into account the different observation times that were used, in order to bring each one into a common standard. This was one of the first research projects that I worked on, with colleagues from the University of East Anglia in the UK, and from the National Oceanic and Atmospheric Administration in Boulder, Colorado. It was tedious work and drove us all crazy at times, but in the end we were able to produce our first estimate of how global air temperatures had varied since 1850, using the best set of data we could come up with. This was not the first time such a graph had been produced, but it was the first major effort to minimize all the uncertainties in the data. Our project was just for land areas, but other colleagues were working hard on the ships' data, too. Before long a comprehensive picture of global temperatures emerged for both land and ocean areas (see figure 2).

Since then, further corrections have been made to the data, other

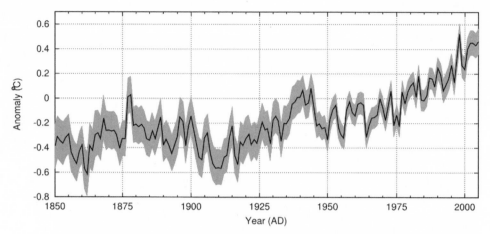

FIGURE 2. Global average annual temperature anomaly time series (°C), known as "HadCRUT3." Anomalies are shown in relation to the average for 1961–1990, which is the zero line on the graph. The black line is the best estimate value; the gray band gives the 95 percent uncertainty range caused by station, sampling, and measurement errors, plus other errors due to limited coverage and biases resulting from urbanization and changes in the exposure of thermometers.

Source: Philip Brohan et al., "Uncertainty Estimates in Regional and Global Observed Temperature Changes: A New Data Set from 1850," Journal of Geophysical Research 111, *(2006): D12106 (doi:10.1029/2005JD006548).*

records have been added, and a comprehensive effort has been made to assess the uncertainties resulting from all these adjustments. The overall picture, however, has not changed much at all. The more recent, highly refined graphs look very much like our earliest tentative efforts. Across the globe, temperatures clearly rose from the late 1800s to the late 1980s (when we were completing our original work). But remarkably, since our first results were published in 1986, temperatures have continued to rise, breaking new records virtually every year. And the pattern is truly global in extent; temperatures were higher in almost every part of the world at the end of the twentieth century than they were at the beginning. This is true for both land and ocean areas.

Given all that, can we say unequivocally that there has there been warming, globally? Absolutely! There is no doubt about it. The average temperature of the atmosphere near the earth's surface has increased by about 0.8°C (1.4°F) over the last 150 years or so; all parts of the earth have become warmer, and the warming has been especially pronounced in recent decades. As discussed earlier, the hockey stick graph shows that these changes are unprecedented for at least a thousand years.

Now, let's be skeptical for a moment. Can the problems we encountered with the instrumental data account for the warming? There is no chance of that at all. How can I be so confident? Because it's not hard to find many independent lines of evidence that fit or confirm the picture we have obtained from the temperature measurements alone.

Perhaps most impressive is the slow but steady rise in sea level that has been observed at tidal stations all over the world. As the oceans warm, they expand. This is a basic physical property of water: you warm it up, and it occupies a slightly larger volume than it did before. As ocean temperatures have risen, sea level has encroached on many low-lying coastal regions throughout the world. This effect has been amplified by melting glaciers. As temperatures have increased, glaciers in high mountain and polar regions have lost mass, and the water from this melting ice has been returned to the world's oceans through streams and rivers, adding to the overall rise in sea level. Over the last hundred years, this additional water has caused the sea level to rise—on average across the globe—by about twenty-five centimeters (ten inches). Glaciers aren't the only things that have been melting. Sea ice in the Arctic has steadily decreased in both thickness and area since satellite records began some thirty years ago. And permafrost— ironically defined as "permanently" frozen ground—has been steadily warming too, resulting in the melting of vast areas of formerly frozen soil across much of Alaska and Siberia, disrupting structures such as roads, houses, and pipelines. All of these factors are clearly related to rising temperatures. Even those people who insist on burying their heads in the sand on the issue of global warming must surely accept that the ground is getting warmer, as we can document from numerous borehole temperature measurements made all over the world.

The biological world is equally aware of these changes in temperature. Plants in many regions are flowering earlier in the spring, and spring migrants—birds, insects, even some migratory fish—have moved farther north (in the Northern Hemisphere) as temperatures have increased. Growing seasons have lengthened as the last spring frosts occur earlier each year, and the first fall frosts arrive later. This has had direct effects on both natural ecosystems and our agricultural systems, allowing certain crops to be grown in ever more northern latitudes, while others are no longer viable where they used to thrive. England now has vineyards in almost all of its many regions, and in southeastern England one eternal optimist has even planted olive trees. In mountain regions, plants have shifted their distributions upslope, literally to cooler pastures, as temperatures have risen in upland areas. In the western United States, a devastating downside of these changes has arisen because a particularly voracious species of tree-killing pine beetle is now able to overwinter in many areas where it used to be killed off by low winter temperatures. With warmer climatic conditions, the beetles can thrive year-round, with the result that over 8 million acres of forest have been destroyed. To make matters worse, the "standing dead" that the beetles leave behind have become a tinderbox, serving as ready fuel for extensive wildfires as warmer and drier conditions compound the situation even further.

To the naysayers and contrarians, all of this is easily explained. The temperature records are biased by "urban heat island effects," meaning that the growth of cities has relentlessly driven temperatures higher. Of course this is true for many locations. But the fact is, climatologists already thought of that. We took it into account by looking at remote rural sites where such effects have not occurred. And temperatures there, too, have gone up and up. Furthermore, sea surface temperatures, as well as temperatures in remote island locations, have risen just as much, so the increase can't be related to urban growth.

Faced with such evidence, denialists typically turn to another specious argument: There's a place that has actually *not* been warming! And there are even a few glaciers that have advanced! This is rather like a person covered in chicken pox pointing to a small patch of skin that does not itch and saying that he isn't really infected. There may indeed be a few places where, for peculiar local reasons, temperatures have not increased, or increases in snowfall have outweighed the effects of warmer temperature on glaciers. But global warming really does mean "global." Around the globe as a whole, temperatures have demonstrably risen in almost every region, with only a few scattered exceptions here and there.

Many denialists also dispute the evidence for sea-level rise; they argue that it's not the sea that is going up but the land that is going down. This is true in many areas—in New Orleans, for example. In fact, in some areas sea level is actually falling as the land surface continues to rise after being depressed, thousands of years ago, by the weight of large ice sheets. In coastal New England and northern Scandinavia, for instance, uplift of the land outpaces sea-level rise, so sea level is falling in those areas, relatively speaking, but it's just a local effect. Once again, *on a global basis,* the evidence for sea-level rise is clear and indisputable. It has even been carefully measured from satellites that can detect changes down to a few millimeters.

But the naysayers persist. Changes in sea ice are also subject to dispute. The records are deemed to be too short, and they note that the ice around Antarctica hasn't declined in area as it has in the Arctic; in fact it may even have increased slightly over the same period. Actually, there is no surprise here; the Southern Hemisphere is mainly oceanic, meaning that changes around Antarctica occur much more slowly than in the Northern Hemisphere.

How naysayers account for the widespread biological changes, the migration of plants and animals, the devastation of forests by pine beetles throughout the western United States all the way up into Alaska, and the upward shift of vegetation belts in mountainous regions is beyond me. It is so abundantly clear that the world is changing a great deal—in both its physical and biological structure, on land and in the sea—that those who willfully ignore the evidence cannot be taken at all seriously.

We can conclude from this discussion that there is a plethora of evidence showing the earth has been getting warmer over the last 150 years, and that the rate of warming has increased in recent decades. So what? Just because temperatures have gone up doesn't necessarily mean that humans are responsible. Isn't it really just part of a natural cycle of "climate change"? Why would these changes have anything to do with human activities? Let's step back from the political spin and consider some basic facts.

The Energy Balance of the Earth

Anything that is warm (meaning above absolute zero, which is very cold, -273°C) radiates energy. The sun, being very hot, radiates a lot of energy, some of which is intercepted by the earth as it revolves around the sun. This energy is absorbed by the earth's surface, which then radiates energy back into space.

Calculations show that if the earth had no atmosphere, the balance of energy received from the sun minus that emitted from the earth would result in a very low surface temperature—around -15°C. But the earth does have an atmosphere—a shallow layer of gases through which the energy from the sun, and the energy radiated away from the earth, must pass. Very hot bodies emit energy at short wavelengths, and cooler bodies emit energy at longer wavelengths, so the energy from the sun is at a much shorter wavelength than that from the earth.

The earth's atmosphere is quite transparent to the shortwave radiation received from the sun, so the sun's energy passes relatively freely through to the earth's surface. The atmosphere, however, is somewhat opaque to the longer-wavelength energy emitted by the earth. That is, some of the energy that the earth emits is absorbed (retained) by the atmospheric gases, thereby warming the atmosphere, and this in turn radiates energy back to the earth in a process that ultimately raises the temperature of both the earth and the atmosphere. Over a long period of time, a balance is reached between the energy received from the sun and the energy radiated from the earth, and this "radiative equilibrium" results in a surface temperature of about 15°C. Because the atmosphere is warmed by the earth—literally from the bottom up—as you leave the earth's surface and ascend, in a balloon or plane, for example, temperature rapidly falls. No surprise, then, that glaciers are found in high mountains, where it is coldest and snow doesn't melt.

Not all gases that make up the mix we call "the atmosphere" are important in this process. It is only certain gases that are found in small quantities—so-called trace gases—that are important. These are carbon dioxide (CO_2), methane (CH_4), nitrous oxide (N_2O), and water vapor (H_2O). Collectively these are referred to as greenhouse gases. This is actually a poor term, as the analogy with heat being trapped inside a greenhouse is not very appropriate; greenhouses heat up mainly because of lack of air circulation. Greenhouse gases absorb the longer-wavelength radiation emitted from the earth, raising the temperature of the atmosphere and increasing its capacity to warm the earth's surface. Of course there are complications in this simple picture, most significantly that clouds block incoming solar energy and reflect energy emitted from the earth's surface. And clouds aren't all the same, so they have different properties of reflection and absorption. But if we look at the big picture, it is the earth's greenhouse gases that make the place habitable by raising temperatures to the levels that we all know and love. Not surprisingly, therefore, the temperature of the earth is particularly sensitive to changes in the concentration of greenhouse gases in the atmosphere. And if those changes occur rapidly, temperatures on earth will quickly follow suit.

All living things on earth have evolved with a certain expectation of what the atmospheric composition is, including the concentrations of greenhouse gases. Yes, these have varied in the past—as I'll explain shortly—but those variations have occurred quite slowly, so living things have had time to adapt. But as far as we can tell, the changes that are happening today are taking place much faster than ever before, so that in the 250-year lifespan of a single long-lived oak tree, for example, the atmosphere in which it has grown since it was just a tiny acorn has changed significantly. Even in my comparatively short career of thirty years or so, carbon dioxide levels have increased more than 25 percent. Although I've given a lot of lectures in that time, it cannot *all* be due to my talking.

To those who have few concerns about the effects of rising levels of carbon dioxide on climate, carbon dioxide is "just plant food." It is the basis of our food chain, because plants absorb it to make leaves, which we then eat. Or we eat other creatures that also eat leaves. So how can more carbon dioxide be bad? It is essential for life on earth, so, surely, more of it can only be good, right? Why should we be concerned about an increase in carbon dioxide and other greenhouse gases?

The problem is that human activity over the past two hundred years or so has dramatically increased the concentration of greenhouse gases in the atmosphere. And "dramatically" is not too strong a word; carbon dioxide levels all over the world have increased by almost 40 percent since around 1800. Furthermore, methane (produced by farm animals and irrigated wetlands such as rice paddies) has more than doubled, and nitrous oxide (from agricultural fertilizers) has gone up by about 15 percent. We have also created entirely new gases, such as chlorofluorocarbons used in industrial processes—substances that never existed in the natural world—and many of these are also greenhouse gases that contribute to further warming of the atmosphere.

As we'll see shortly, there is no evidence that such rapid changes occurred at any time in at least the last 850,000 years. That figure takes us back to an age when early humans were just learning how to make fire. Changes in the composition of the atmosphere, happening so quickly, result in similarly rapid changes in the earth's climate, and that's where the problem lies. Life on earth has evolved in balance with all the natural processes that occur on earth, and for the most part these have changed relatively slowly. The dramatic atmospheric changes of recent years have disrupted the natural balance, and it's this balance, this expected set of conditions, that we and other members of the earth's living communities have adapted to. That's what we mean when we speak about the climate of the region where we live.

Your Climate Is What You're Used To

The energy that the earth receives from the sun is not distributed equally. Because the earth is a sphere that rotates on its axis as it revolves around the sun, more energy is received near the Equator and in the tropics than at higher latitudes, and more is received (in each hemisphere) during summer months than in winter months. These factors are what cause the atmosphere and the oceans to circulate, redistributing the energy around the globe. Continents and mountain ranges get in the way, forcing ocean currents to carry warm (or cold) water to places where they might not otherwise go, and causing the atmosphere to swerve far to the north or south as it sweeps across the globe. The immediate consequences of these processes are what we think of as weather—the daily and seasonal variations in temperature, rainfall, humidity, cloudiness, and so on that characterize each region. But over time, the same kinds of weather events tend to recur within each region—of course, a bit mixed up from one year to the next. This general repetition, season to season, year to year, gives each place its distinct "climate."

I was thinking about this as I wandered around Plimoth Plantation in Plymouth, Massachusetts. This "1627 English Village" is a modern re-creation of what life must have been like for the early colonists when they arrived in Massachusetts centuries ago. When the Pilgrims left England in the early seventeenth century to settle the New World, they took with them more than the basic necessities for founding a colony. They also took their collective experiences of life in England—including English weather. Having grown up on the Atlantic edge of Europe, they had built up a mental picture of what the weather in their villages and towns was like. In effect, they had accumulated a large sample of "weather experiences" that would have enabled them to describe what a typical winter or summer might be like "back home," what extremes they would expect, and so forth. Their large sample of weather experiences provided a mental impression of the everyday weather events that over a long enough period of time would have included the full range of conditions likely to be experienced in their villages and towns.

When the Pilgrims arrived in "New" England, the experience of weather in their strange new world was quite different: higher temperatures and more humid conditions with thunderstorms quite common in the summer, and much colder, snowier winters, causing lakes and ponds to freeze. Over time, they may have experienced occasional hurricanes, or the severe winter storms known as nor'easters, and perhaps even a rare tornado. It would take many years before these experiences would provide them with a mental picture of the climate of New England to compare with that of the English climate they'd left behind.

In each region, plants and animals (including insects and people) have become adapted to this experience we call climate. Ecosystems in each region have evolved to the point where they are optimized for those particular climatic conditions. Soils have developed over thousands of years as a result of the interactions between the climate, vegetation, and the underlying rocks. Agricultural systems have been developed to maximize the yield of specific crops. We know what will grow, what to wear, what kind of housing we need in a particular climate, and we have built complicated and expensive infrastructures to meet those needs. For example, many cities and towns have invested in fleets of snow removal vehicles to maintain clear highways in winter. In low-lying areas, floodplain and coastal management systems have been adopted, and communities have been constructed to deal with high water by using appropriate devices such as culverts and storm drains. In other areas, farmers have spent millions on irrigation systems to reduce the effects of prolonged dry spells and to maximize crop production. All of these systems rely to a large extent on this collective experience of weather—our climate—remaining relatively stable.

If climate changes slowly (as it certainly has over long periods of time), the changes can generally be managed in each region. The various plants and animals that make up an ecosystem may slowly migrate to new areas, and those components may become reorganized as they migrate, creating new ecosystems unlike those that existed before. These changes take place over many centuries, perhaps even thousands of years. Such changes are quite normal. But if change happens too fast, things can go wrong. If you trip and fall down, you won't hurt yourself too badly; your system is capable of handling the speed at which you hit the ground. But if you fall from a ten-story building, there's likely to be a different result. Your system is not adapted to deal with the much faster rate of descent in the second case. In the same way, all systems have evolved to cope with the normal variability in existing environmental conditions. If things change too quickly, too unexpectedly, a system may not be able to adapt to the new conditions, and the result will be disruption of that system.

How does this apply to the greenhouse gas issue? One of the difficulties we have as short-lived human beings is that our perspective on changes in the earth's major systems (the atmosphere, oceans, the world of ice, the biosphere, and the land surface) is quite limited. *Homo sapiens,* our noble species, has been around for only about 130,000 years. For most of that time we were hunters and gatherers, wandering the earth in pursuit of our basic needs. Agriculture put a stop to most of those wanderings, permitting a more settled, sedentary way of life. But agriculture began only around eleven thousand years ago. The first real city-like settlements arose about five thousand to six thousand years ago. Most of our written history spans just the last two thousand to three thousand years. Modern scientific investigations got under way only in the last few hundred years, and measurements of our weather extend back just one or two centuries in most parts of the world. In remote places such as Antarctica and the Arctic, weather records are barely fifty years in length. And the same goes for instrumental measurements of carbon dioxide. The first measurement program to monitor carbon dioxide was begun by Charles Keeling in 1957 on Mauna Loa, Hawaii, and that data set gives us the longest instrumental record that we have on changes in the levels of this important greenhouse gas. All of these dates and numbers mean that, in terms of the earth's long and eventful history, our perspective on changes in the environment is extremely limited. Thanks to a lot of very bright and creative people, however, it has been possible to extend that perspective much further back in time.

In the case of carbon dioxide and other greenhouse gases, a very long perspective on atmospheric concentrations has been possible because tiny samples of the atmosphere from times past are trapped in bubbles buried deep beneath the snow on the Antarctic ice sheet. I like to imagine that somebody once dropped a chunk of ice chipped off a glacier somewhere into a glass of scotch, and watched the bubbles rapidly explode into the liquid as the ice melted and "de-gassed." No doubt as the thoughtful tippler sipped away, he mused to himself, "Hmm, I wonder if I can use this phenomenon to reconstruct the history of atmospheric greenhouse gases?"

Scotch or no scotch, the idea did occur to Hans Oeschger, a physicist from Bern, Switzerland. Hans was working with a team of Danish and U.S. scientists who in 1981 had recovered a core of ice by drilling right through the Greenland ice sheet. The goal was to obtain a long climatic record by studying chemical changes in snow that had accumulated on the ice sheet over tens of thousands of years. This idea had come about through research started by Willi Dansgaard, from Copenhagen. Dansgaard had been studying isotopes in rainwater. Isotopes are elements with a slightly different atomic mass. Different combinations of the isotopes of oxygen and hydrogen in water produce water molecules with different physical properties, and these can be used to figure out what the temperature was when a droplet of water (or a snowflake) condensed in a cloud. He was studying this phenomenon near his home in Denmark by collecting rain and snow as storms passed overhead. One day while doing this research he had a startling idea: if you could drill down through the Greenland ice sheet and recover a core of ice (that is, a sample made up of all the layers of snow that had fallen over time), you might be able to obtain a long record of what the temperature had been in the atmosphere above the ice sheet when the snowflakes formed. In effect, you would have an "atmospheric thermometer" stretching far back in time, depending on how deep you drilled into the ice sheet. To make this idea work, you'd have to go to the highest part of the Greenland ice sheet, where there is virtually no melting, so that each new layer of snow simply buries the previous layer, building up layer on layer, as seen with rings of wood in a tree stump.

To accomplish his goal, Dansgaard teamed up with Chet Langway, from the State University of New York in Buffalo, who led a U.S. research group with experience in drilling long cores of ice from the Greenland ice sheet. They first recovered a core 1,370 meters long (about 4,500 feet) from Camp Century, not far from the Thule U.S. military airbase in northwestern Greenland. Analysis of this ice core nicely demonstrated the principle that Dansgaard had postulated back in Denmark: the isotopes of oxygen in the ice did indeed give a remarkable picture of how temperatures had varied in northern Greenland over tens of thousands of years. But the ice core site was not ideal, as it posed difficulties in figuring out just how old the ice was as the researchers drilled into the ice sheet. Langway, Dansgaard, and Oeschger decided to team up to drill a second core from a site in southern Greenland (called Dye 3) where the ice layers were easier to interpret. This time the core extended right to the bottom of the ice sheet, about three kilometers (roughly ten thousand feet) below the surface.

The prospect of obtaining a complete record of both atmospheric temperature and carbon dioxide from this long ice core was really exciting. But as it turned out, Hans Oeschger's attempts to get a history of atmospheric carbon dioxide from the gas bubbles in that ice were not entirely successful, in part because of contamination of the ice by calcareous dust and surface melting. Nevertheless, the experiment led to other efforts using ice cores from the opposite side of the earth (Antarctica), where it was much colder and contamination was not a problem. Carbon dioxide mixes very quickly in the atmosphere, so whether you measure the atmospheric concentration in Greenland or in Antarctica, the figures are almost the same on an annual basis. In Antarctica, the research was led by French scientists Claude Lorius, Jean Jouzel, and Dominique Raynaud using ice from the Russian research station, Vostok. The Vostok ice core, which extended back about 450,000 years, provided a truly remarkable long-term view of past climate and changes in greenhouse gases (see figure 3). For the first time, it was possible to see a clear pattern of changes in air temperature (from hydrogen isotopes in the ice) and trace gases (CO_2, CH_4) from bubbles of air trapped as the snow became compressed into ice.

The Vostok ice core showed that temperatures and greenhouse gases had varied in lockstep; more important, it showed that there were natural limits to the levels of greenhouse gases in the atmosphere. Over the entire period, carbon dioxide had never gone below 180 parts per million (ppm) or above 280 ppm. Some would argue

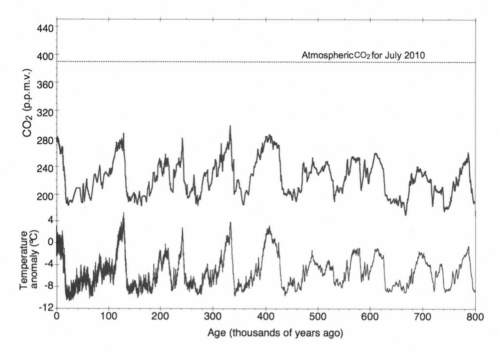

FIGURE 3. (top graph) Atmospheric carbon dioxide levels obtained from bubbles trapped in an ice core from Vostok in Antarctica (measured in parts per million by volume, or ppmv). The record extends back (from left to right) from the pre-industrial period to 800,000 years ago. Carbon dioxide levels reached 390ppmv in 2010.

(bottom graph) Estimates of temperature differences from today at Vostok, based on deuterium isotopes in the ice.

Source: Adapted from Dieter Lüthi et al., "High-Resolution Carbon Dioxide Concentration Record 650,000–800,000 Years before Present," Nature 453 (2008): 379–82.

that these limits demonstrate that the earth has a natural "balance"; when the pendulum swings too far in one direction, various processes bring it back toward the middle so that the atmospheric composition stays within certain seemingly fixed limits. In fact, since the Vostok ice core was first studied, a new and even longer core has been recovered from another part of Antarctica, and this extends even further back in time—to over 850,000 years. Remarkably, this core also shows

CHAPTER FIVE

that over the entire period—more than four times the length of time our species has been on earth—the carbon dioxide levels have always stayed between those same upper and lower limits of 280 and 180 ppm, respectively, dropping down during the ice ages and increasing during warmer intervals.

The swings in temperature, leading to the growth of large ice sheets on continents in the Northern Hemisphere, were the result of a slow shift in the position of the earth in relation to the sun. These changes, resulting from very small gravitational effects of the planets on the earth's orbit, were figured out by a Serbian mathematician, Milutin Milankovic, in the 1930s. Nowadays we recognize that Milankovic's "orbital theory of glaciation" provides quite a robust explanation for most of the very long-term, slow changes over the earth's history. But the important point here is that these are *s-l-o-w* changes, such that an entire cycle from a warm interglacial to a glacial period took about 100,000 years.

We are comfortably enjoying the latest interglacial period now, so we might expect that carbon dioxide levels would be just like those in all previous interglacials: about 280 ppm. In fact that is approximately what carbon dioxide levels were before the mid-eighteenth century. Then along came James Watt. It was Watt who, in 1784, invented the steam engine, which led to the Industrial Revolution.

Steam engines were fueled by coal, so as more and more of them came into use, they created an explosive demand for that shiny black rock, to power factories and trains. Coal consists primarily of compressed plant material, much of it dating back to the Carboniferous period about 330 million years ago. At that time the carbon dioxide level in the atmosphere was high—much higher than today—but plants extracted the carbon dioxide from the atmosphere (through the process of photosynthesis) and converted it into carbohydrates in their leaves. And as they did so, the level of carbon dioxide in the atmosphere declined. Over eons of geological time, those plants became compressed into the coal deposits that we see today. By burning them in factories, fireplaces, and steam engines, we returned that "old" carbon dioxide to the atmosphere, and as the Industrial Revolution

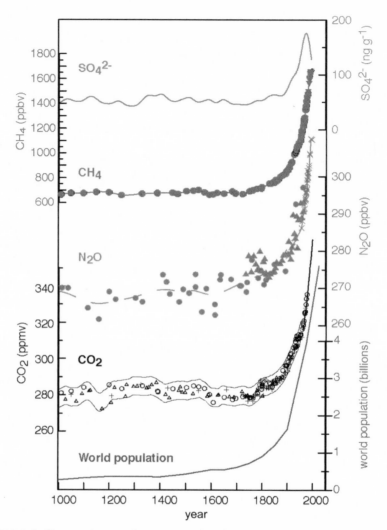

FIGURE 4. Changes in greenhouse gases (methane, CH₄, nitrous oxide, N₂O) and carbon dioxide (CO_2) over the past one thousand years, from gas bubbles in an ice core from Greenland. The overall rise in greenhouse gases parallels world population growth (bottom line). The abrupt increase in carbon dioxide corresponds to the onset of the Industrial Revolution in the early nineteenth century. The top line shows changes in sulfate, which reflects air pollution over North America. The recent decline is the result of the Clean Air Act, which significantly reduced air pollution even as far away as Greenland.

Source: Dominique Raynaud et al., "The Late Quaternary History of Atmospheric Trace Gases and Aerosols: Interactions between Climate and Biogeochemical Cycles," in Paleoclimate, Global Change and the Future, *ed. K. D. Alverson, R. S. Bradley and T. F. Pedersen (Berlin: Springer, 2003), 13–31. With kind permission of Springer Science and Business Media*

gained steam (so to speak), the level of carbon dioxide in the atmosphere began to rise quickly. Simply stated, carbon dioxide was going back into the atmosphere thousands of time faster than the rate at which it had been removed from the atmosphere long ago.

A century or so after Watt's invention, Gustav Daimler and Karl Benz patented another earthshaking invention, the internal combustion engine. This invention did not run on coal but instead used petroleum extracted from oil, which derives from microscopic photosynthetic organisms (and the tiny animals that lived on them) that removed carbon dioxide from the atmosphere tens of millions of years ago. Over eons of geological time, compression of the deep ocean sediments that were almost entirely made up of these organisms resulted in the formation of oil and gas deposits. As with the combustion of coal, by burning oil and gas, we are putting back into the atmosphere carbon dioxide that was removed tens of millions of years ago. As the demand for petroleum products has increased, so has the amount of old carbon being returned to the atmosphere (see figure 4).

All this combustion of coal, oil, and gas has been driven by the explosive growth in world population and the ever-increasing demand for energy-consuming products: cars, refrigerators, lighting, computers, TV sets, and a host of other items. If we add up (or at least estimate) how many people lived on earth throughout history up to the year 1800, the figure is probably fewer than 1 billion people *in total*. Between 1800 and 1930, another 1 billion people were born. Nowadays we add 1 billion more people *every 12 or 13 years,* so the earth's population currently exceeds 6.5 billion. And the United Nations estimates that by later in this century it will be over 9 billion.

Of course this population increase has not been uniform across the earth, and it will not be uniform going forward in time, either. Most of the population growth has occurred in the developing world, and that will be so for the rest of this century. Those billions, many of whom lead extremely difficult lives, would very much like to have the facilities and products that most of us in the developed world take for granted. As their numbers grow and demand inevitably increases, so does the consumption of fossil fuel—with direct effects on the concentration of trace gases in the atmosphere. Today, carbon dioxide levels have reached 390 ppm, almost 40 percent higher than the levels that were common before the first cities were built, before writing was invented, even before the Industrial Revolution took place. That increase, in just 250 years, far exceeds the rate of change that we have seen in the ice core records. It greatly exceeds any "natural" process that the earth has witnessed in at least the last 850,000 years. And it has brought the atmospheric carbon dioxide concentration to a level far above that experienced by any of our ancestors, going all the way back to the hunters and gatherers who inhabited Olduvai Gorge a million years ago.

Unfortunately, the amount of carbon dioxide in the atmosphere is likely to rise to an even higher level, and the rate of this increase may even accelerate. This is a simple function of world population growth and an ever-rising standard of living. For the world's poorest people, improvements to their way of life can't come quickly enough, and those improvements almost always involve an increase in energy use. If world population increases from approximately 6 billion today to 9 billion later in the twenty-first century, even more of the "old" carbon that was extracted from the atmosphere eons ago will be returned to the atmosphere as we burn more and more fossil fuel—to generate electricity; to heat or cool our homes, schools, shops, and workplaces; and to propel our vehicles. At least that is the dismal prospect that we face if nothing changes—the so-called business-as-usual scenario. There are alternatives, though, which I discuss later.

Does it really matter if carbon dioxide increases that much? Carbon dioxide levels were a good deal higher in the geological past, so why should we care if it goes back to that situation again? If we take an objective, long-term view of the earth—perhaps the view that another civilization in another solar system might have—no, it really doesn't matter. Planet Earth will survive; of course, it may be uninhabitable (and eventually uninhabited), but if you take a geological perspective—looking at how things might change over the next million years or two—the "industrial age" will be just a minor blip in time. Nevertheless, we, the current residents of Planet Earth, don't have the luxury of that perspective; the timescale on which we experience change is so much shorter—seasons, years, decades—that changes which might seem minor on a very long scale are much more immediate for us, taking place here and now, during our lifetime and the lifetimes of our children. Furthermore, we have so altered the earth's surface—through agriculture, road building, destruction of natural vegetation, damming and diversion of waterways, and so on—that other animals (and even plants) can no longer simply migrate to keep pace with changes in climate, as many undoubtedly did in the past. "Habitat tracking" was a critical evolutionary strategy, so that as climates shifted as a result of completely natural forces in times past, plants and animals, including human beings, generally survived those disruptions. Today's rate of change, though, is much more rapid than anything we know about from past history.

No ice core or other geological record provides any evidence for a global-scale doubling or tripling of greenhouse gas concentrations within a few centuries, as may well turn out to be the case, from about 1800 to 2100. So it is the rate and magnitude of the greenhouse gas increases, together with the burgeoning global population and the destruction of the natural environment (particularly tropical forests, which are efficient at removing carbon dioxide from the air), that is the cause for concern. Any one of these factors would be a big problem. Altogether they create a need, in fact an obligation, to reconsider past strategies for growth and energy use completely, in order to set a new direction—toward a sustainable future.

6

Climate Futures
Where Are We Heading?

Some things you just can't keep to yourself—you have to share. So it is with the atmosphere; it blows around the world, and we all breathe it. Whatever we put into it gets passed on, whether the next person wants it or not. This has always been a problem in places such as western Europe, where small countries, closely juxtaposed, may receive pollutants from whatever power plant or industrial polluter happens to be upwind. Even in faraway Greenland, pollutants from the Industrial Revolution can be clearly detected in the otherwise pristine snows of the far North.

Dealing with these problems has led to a lot of legal disputes. New England states sued power plants in Ohio for putting sulfur dioxide into the air, which ended up as sulfuric acid, acidifying lakes and streams throughout the Northeast. Eventually the federal government regulated emissions through the Clean Air Act, which has been quite effective in reducing air pollution. You can even see the downturn in sulfur deposition far away in Greenland ice cores (see figure 4 in chapter 5). The mechanism for reducing sulfur dioxide was "cap and trade." This means that limits were placed on total emissions, and power plants had to clean up their act or else purchase pollution "credits" from other, cleaner plants. Slowly the limits were lowered so that it became economically more attractive to install scrubbers and other antipollution devices rather than buying credits. The system gradually made industry cleaner by using the "marketplace" as a tool to promote less noxious emissions.

The world has experience, then, in dealing with atmospheric pollution problems on a regional scale. The problem was recognized, a solution was devised and implemented, and the problem was mitigated. Dealing with greenhouse gases, and carbon dioxide in particular, is a much more difficult matter. The high standard of living enjoyed by modern society is built on the combustion of fossil fuel—in generating electricity, in heating or cooling homes and factories, and in transportation, via cars, trains, planes, buses, and ships. Changing our consumption of the energy resources that were created long ago, in distant geological time, is an immense challenge. But unless we do so, we face difficult times ahead.

This threat was recognized in a remarkably prescient document, the "United Nations Framework Convention on Climate Change," which was prepared in 1992. It is a wonderful thing to read, full of diplomatic phrases that remind me of conversations old men might have in a private gentlemen's club in London. It begins:

> *The Parties to this Convention,*
>
> *Acknowledging* that change in the Earth's climate and its adverse effects are a common concern of humankind . . .
>
> *Concerned* that human activities have been substantially increasing the atmospheric concentrations of greenhouse gases . . .
>
> *Noting* that the largest share of historical and current global emissions of greenhouse gases has originated in developed countries . . .
>
> *Aware* of the role and importance in terrestrial and marine ecosystems of sinks and reservoirs of greenhouse gases . . .

And so it goes on, "*noting . . . recalling . . . recalling also . . . recalling further . . . recognizing . . .* and *reaffirming*," until it reaches its magnificent finale:

> *Determined* [yes, determined, dammit!] to protect the climate system for present and future generations,
>
> *The Parties to this Convention have agreed* [finally!] *as follows:*
>
> . . . to achieve . . . stabilization of greenhouse gas concentrations in the atmosphere at a level that would prevent dangerous anthropogenic interference with the climate system. Such a level should

be achieved within a time-frame sufficient to allow ecosystems to adapt naturally to climate change, to ensure that food production is not threatened and to enable economic development to proceed in a sustainable manner.[1]

By 2010, virtually every card-carrying member of the United Nations—all 192 countries—had ratified this convention, clearly recognizing the common atmospheric resource that we all share, and the potential for danger in the future if we are not careful. But the wording of this document is pretty vague. What exactly does "dangerous anthropogenic interference with the climate system" mean? Who decides what is dangerous? And how long is "a time-frame sufficient to allow ecosystems to adapt naturally"?

For most people, daily life poses few dangers. The most hazardous thing we do is to drive on a busy highway, sometimes faster than we should. But experience tells us that if we feel we are going too fast, all we have to do is ease off on the accelerator and we will slow down. If we take our foot off the pedal completely, we will eventually come to a stop: speeding problem solved. It is not surprising, then, that most people think that if we decide we (that is, the world community) are using too much fossil fuel and producing too much carbon dioxide, we can resolve the problem simply by reducing emissions, quickly lowering carbon dioxide levels in the atmosphere. Unfortunately, this does not work. Carbon dioxide is a very persistent gas. It will be a very long time before the CO_2 we produce today is removed from the atmosphere. The amount of carbon dioxide in the atmosphere is the result of a balance between sources and sinks, that is, between the processes that produce CO_2 and those that absorb it. In adding carbon dioxide to the atmosphere by burning fossil fuels, we have disturbed the balance between production and loss that occurred in the natural world, and it will take a long time for those natural processes to "catch up" with this disturbance and restore a new balance to the global system.

This is not as true for the other greenhouse gases—methane and nitrous oxide, for example—which have a relatively short life in the atmosphere. But carbon dioxide is different, and therein lies a very big problem. Taking our collective foot off the fossil fuel accelerator will cause the level of carbon dioxide in the atmosphere to drop, but only very slowly. If we continued to increase CO_2 levels to, say, 450 parts per million (a virtual certainty, given that by the beginning of this century we were at 390 ppm) and then suddenly decided to stop all fossil fuel consumption—instantly, overnight—carbon dioxide levels would gradually decline, but they would not drop below about 350 ppm *for the next one thousand years*! Remember that carbon dioxide levels before the Industrial Revolution were around 280 ppm, and that was as high as they had reached in any of the past warm interglacial periods. In other words, carbon dioxide levels would remain, well into the future, far above anything the world has seen for at least the last 850,000 years. Our thinking about "dangerous interference" has to take all of this into account. The more carbon dioxide we put into the atmosphere, the longer its impact will last, and the higher the long-term level of CO_2 in the atmosphere. To put this another way, we can't just keep "testing the waters"—adding more and more carbon dioxide to the atmosphere until things "start to go wrong" with the global climate, and only then put the brakes on. At that point, even completely eliminating fossil fuel use overnight will bring about only a very, very slow decline in the atmospheric concentration, and will do little to rectify the immediate climate problem.

Given that we have increased preindustrial levels of carbon dioxide by about 40 percent, it is simply inevitable that levels are going to remain high for a long time to come. But although carbon dioxide in the atmosphere has increased, atmospheric temperatures have not (yet) gone up as much as we might expect. That is, if we use our knowledge of the physics of the atmosphere, and what we know about how carbon dioxide affects the transfer of energy from the earth's surface back to outer space, it is clear that the warming of the atmosphere observed so far is considerably less than it "should" be. The reason is that much of the heat is being carried away into the oceans, warming the surface waters and then being mixed to ever greater depths. This doesn't happen over land areas because there is no mixing as there is in bodies of water.

In effect, the energy is being banked—stored away in the great oceans of the world. We can see this clearly by looking at the millions of temperature measurements that have been taken by ships and buoys over the last hundred years. Temperatures at the surface of the oceans have risen, but so have temperatures at increasingly greater depths. This means that even if we stopped producing carbon dioxide today—completely—global warming would continue far into the future as the oceans slowly released all the stored heat back into the atmosphere until a new balance was achieved. It follows, therefore, that warm conditions are going to be with us for generations to come. But there is another problem that goes along with this reality.

As noted earlier, when the oceans warm, the volume of water increases, since warm water occupies a slightly larger volume than colder water. You'd never notice this in a glass, but for a body of water the size of the world's oceans, it is quite a significant factor. Thermal expansion of the oceans will increase sea level around the world by at least 0.5 meters (eighteen inches) over the next eighty to ninety years. And this is a persistent effect, because it takes a very long time for the oceans to cool off. So a rising sea level is inevitable, a dead certainty that will result from the warming we have already brought about. This does not even take into account any additional melting from ice caps and glaciers, which are all retreating quite rapidly.

When we add up all these effects, most experts now believe that sea level will rise by at least one meter (about three feet) by the end of the century, and perhaps by as much as two meters (six feet). If there were any sudden collapse of ice shelves, as some predict, the change would be even greater. Today, over 100 million people live close to the ocean, within the coastal zone that is less than one meter above present sea level. Along the coast of the Nile delta in Egypt, 6 million people are vulnerable, and in the coastal lowlands of Bangladesh, the figure is closer to 15 million. Some island nations, such as Kiribati in the Pacific and the Maldives in the Indian Ocean, are entirely flat, with no land more than two meters above sea level. If our current understanding of the problem is correct, there is simply no future for these countries.

There is an additional problem involving the oceans. As I have said, much of the carbon dioxide from the burning of fossil fuel is absorbed by the oceans; otherwise the global warming problem would be far worse than it is. This means that for every ton of carbon dioxide we produce, only about 0.4 tons remains in the atmosphere. The rest dissolves in the oceans. But there is a secondary effect that is a problem all by itself. Carbon dioxide dissolved in water produces a weak acid (carbonic acid), and so the oceans are slowly becoming more acidic. There are very few long-term measurements of ocean acidity, but one of the longest sets of precise measurements (taken north of Hawaii) shows an alarming trend toward increasingly acidic conditions near the surface and at depth. If we continue to use up fossil fuel at the rate we have over the last twenty years or so, the oceans will become more and more acidic. At some point this is likely to become a problem for certain types of marine life, such as corals, which secrete calcareous skeletons to survive. Calcium carbonate shells and acidic waters are not a healthy combination for corals.

All this is depressing enough, but we also have to bear in mind that "global warming" is only one simple measure of the effect of increasing greenhouse gases in the atmosphere. As the energy balance of the earth is changed, so too is the circulation of the atmosphere. The earth's great wind systems are really just a response to the uneven distribution of heat across the globe. And the patterns of rainfall are closely linked to the way in which winds move moisture around the earth. Global warming will alter the pattern of wind systems, and consequently the distribution of rainfall will be altered too. But unlike the expected changes of temperature—which will increase *everywhere* on earth—the changes in rainfall are much more variable. It is much harder to estimate rainfall changes, but figure 5 shows the average change in dry season rainfall, based on twenty-two models of future climate change, assuming greenhouse gases continue to increase at a moderate rate. Although the details may not be perfectly accurate, the overall picture is quite clear: high-latitude regions will become wetter, but many areas in the subtropics and lower latitudes will become much

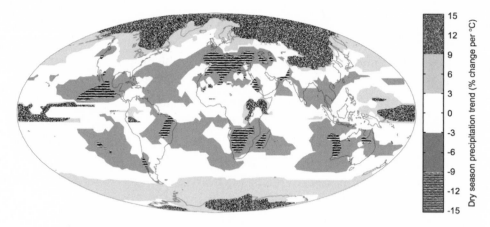

FIGURE 5. Expected changes in dry season rainfall (as a percentage of the average from 1900 to 1950) for every 1°C of warming in a "mid-range" future scenario (A1B; see figure 6), based on the average of 22 global circulation models. All areas that are designated as negative will become drier (mainly the tropics and subtropics), and the positive values will become wetter (mainly at higher latitudes). For example, future summer rainfall amounts in the Mediterranean will be about 10 percent lower for every 1°C of warming, whereas summers in eastern Canada will be wetter by 10 percent or more for every 1°C of warming. White areas show those regions where there is too much disagreement among the models to be confident in the expected change.

Source: Adapted from Susan Solomon et al., "Irreversible Climate Change due to Carbon Dioxide Emissions," Proceedings of the National Academy of Sciences of the U.S. *106 (2009): 1704–9.*

drier. The lesson here is that "global warming" may mean "regional drying" in many important agricultural areas.

With this brief background out of the way, we can now revisit the question of "dangerous anthropogenic interference with the climate system."

On the face of it, deciding what is dangerous is a value judgment that different people may disagree on. For me, leaping off a bridge with a rope tied around my ankle is a dangerous thing to do, but to avid bungee jumpers, it's good clean fun. Similarly, the UN statement leaves plenty of room for discussion and debate. Nevertheless, it does provide some guidance; it aims to achieve stabilization of greenhouse gas concentrations in the atmosphere so that ecosystems will be able to adapt naturally to climate change, ensuring that food production is not threatened and enabling economic development to proceed in a sustainable manner.

Clearly, if sea level rises one to two meters and your entire country is below that level, this will not allow you to develop in a sustainable manner. For tens of millions of people who live and farm along the coast (as in Bangladesh or Egypt), rising sea level promises only social disruption and economic disaster. And the problem is not confined just to developing countries, as we saw quite vividly with the devastation brought about by the flooding of New Orleans from Hurricane Katrina in 2005. Low-lying coastal regions, wherever they may be, are vulnerable to sea-level rise, even without the extra storm surges associated with severe storms. Next time you go down into the subway in New York, Tokyo, Los Angeles, or any number of large metropolitan centers built right on the coast, think about how many meters above sea level that hole in the ground might be. It won't take a big change in sea level to wreak havoc in many places around the globe.

Some issues can readily be brought into realistic focus in discussions of climate change, but other aspects are less clear-cut. At what point will changes of rainfall or temperature threaten food production? How fast is too fast for ecosystems to adapt "naturally"? In many countries, vast numbers of people lead a simple agricultural life in which they rely on whatever they can grow to survive. If their crops fail, the consequences are immediate and dire. But in the wealthier countries, where most people interact with the agricultural system only in the cooler sections of the supermarket, crop failures are not a life-and-death issue. Prices may go up, but lives are not threatened. Farmers may suffer economic losses, but often these are buffered by insurance or government subsidies for "disasters." The reality is that, for most of the developed world, the effects of climate change on food production are well below the radar.

Ironically, this is not the case with the issue of ecosystem adaptation, which is often portrayed in terms of the survival of iconic species. Enter the polar bear, a familiar symbol of global warming. As the planet warms, and the Arctic becomes warmer, sea ice recedes and gets thinner, at which point (it is argued) the polar bears all die. Such arguments led the Environmental Protection Agency to declare the polar bear in Alaska an endangered species, the first mammal to be so listed—a victim of global warming. Passions are easily raised over the endangered polar bears, but so far there has been no comparable campaign to save the lonely Kiribati farmer whose lands are slowly disappearing beneath the waves, or the desperately poor Bangladeshi whose water supply is gradually becoming contaminated by saltwater as the sea encroaches ever higher. It is within this complex set of values and concerns, with cute images at one end of the spectrum and unimaginable poverty at the other, that politicians must decide how best to balance the interests of their constituents with the larger issues of global equity.

The first attempt to address the problem on a global scale was the Kyoto Protocol, a treaty that was designed to reduce greenhouse gas emissions and their attendant effects on global climate. Delegates from all over the world met in Kyoto, Japan, and after much wrangling agreed on the broad outline of an agreement on December 11, 1997. Vice President Al Gore, in a very dramatic move, flew in to pledge that the United States would sign on to the treaty's goals. At that time his office issued the following statement: "U.S. leadership was instrumental in achieving a strong and realistic agreement in Kyoto—one that couples ambitious environmental targets with flexible market mechanisms to meet those goals at the lowest possible cost." This sounded pretty good, but there was a catch: "Signing the Protocol, while an important step forward, imposes no obligations on the United States. The Protocol becomes binding only with the advice and consent of the U.S. Senate. As we have said before, we will not submit the Protocol for ratification without the meaningful participation of key developing countries in efforts to address climate change."[2] He might have added, "Therefore this is a futile and meaningless gesture, because I know damn well that the Senate will never pass it."

U.S. politics being what it is, reaction was swift. Within a few months, ninety-five senators adopted a resolution stating that "the United States should not be a signatory to any protocol to, or other agreement regarding, the United Nations Framework Convention on Climate Change of 1992, at negotiations in Kyoto." Not one senator opposed the resolution. Senator James Inhofe, never shy on the topic, summarized the consensus thus: "The Kyoto Protocol is a lot of economic pain for no climate gain."[3]

In some ways you could argue that he was right: any reduction in carbon dioxide would have very little *immediate* effect on global warming because of the built-in warming that is already stored in the oceans, and because carbon dioxide is only very slowly removed from the atmosphere. But Inhofe's argument misses the most important point: the longer you wait to control emissions, the worse the problem becomes, while the sooner you tackle the issue, the less the long-term impact. Furthermore, distinguished economists who have examined the problem clearly show that the long-range costs of doing nothing are far greater than the costs of taking the problem on right now. In effect, they argue that we can't afford *not* to take action right away. But this is not an argument that is very attractive to a politician like Inhofe. You are asking somebody who has to run for reelection every few years to pass legislation that may have a negative impact on his constituents in the short term (perhaps through higher energy costs), even though there may be long-term benefits to their children and grandchildren. But unfortunately for them, the kids can't vote.

And so in the United States, Kyoto went down in flames. Gore remained optimistic, issuing a statement on November 12, 1998, in which he said, "We are confident that in time the nations of the world will arrive at a course that maintains strong and sustainable economic growth, respects the needs and aspirations of all nations, and protects future generations from the threat of global warming."[4]

In fact, most other nations of the world have indeed been far more receptive to the Kyoto Protocol than the United States. By 2010, virtually all the other countries (183 in all) had signed on. Only the United States has decreed that it will not.

So what, exactly, did these countries agree to in this treaty?

The Kyoto Protocol identified four greenhouse gases—carbon dioxide, methane, nitrous oxide, and sulfur hexafluoride—plus some other industrial gases that needed to be reduced. The industrialized countries agreed collectively to reduce emissions 5.2 percent below 1990 levels. No limits, however, were placed on developing countries such as China and India. This omission became the major bone of contention in the U.S. Senate. Why should China, which today produces as much carbon dioxide as the United States, be exempt from limiting its emissions? The logic, in fact, is simple. The Industrial Revolution began in the UK, then spread to the rest of the world, and the fuel that drove this revolution was coal. Later, the development of the automobile led to a surge in demand for oil. The combustion of billions of tons of coal and billions of barrels of oil since the eighteenth century has enabled (mainly) Western countries to thrive, raising the standard of living for hundreds of millions of people. But the overall effect of that growth and prosperity has been a steady accumulation of greenhouse gases in the atmosphere. More than 73 percent of the "extra" carbon dioxide in the atmosphere (that is, the amount over and above the natural background level) has been produced by the industrialized countries over the last 250 years. So, logically, it is their responsibility to clean up the problem they have largely created. To India and China (and other developing countries such as Indonesia and Brazil), asking them to reduce their emissions of carbon dioxide now is tantamount to asking them to limit improvements to their standard of living, which is well below that of the developed world. From their perspective, it's a question of "you broke it, you fix it."

I think of this problem as rather like a long wall covered with graffiti. People have been writing on the wall for years, and it looks a mess. Then along comes a Chinese guy and an Indian, and they both spray something on the wall—just a few characters. Suddenly, all the other people who messed up the wall in the first place grab them by the arms and say, "Hey—it's time we all cleaned this up; let's just divide the wall into sections, and we'll all take equal responsibility for the clean-up." Clearly, this is not very fair.

Fairness and logic, however, do not always play well in the U.S. Congress. I recall a conversation I had with Congressman Jim Sensenbrenner, a Wisconsin Republican, who accompanied me and a group of other scientists, as well as the Norwegian science minister, on a trip to inspect research bases in the Norwegian Arctic. Back in the bar of the hotel, I seized the rare opportunity to talk to an influential congressman in a casual setting.

"So," I asked, "how come the United States won't ratify the Kyoto Protocol?"

Sensenbrenner visibly stiffened. "Kyoto is a conspiracy," he replied. "Countries like China, Brazil, India, and Indonesia can't compete with the United States. This is a mechanism they have devised to cripple the U.S. economy so the playing field will tilt in their favor."

Now, I had thought of many reasons why someone might not support Kyoto, but the notion that it was all a vast international conspiracy to destroy the U.S. economy was something I really had not considered. I wondered how many other conspiracies I had overlooked. How could I have been so naïve? But on reflection (which took a millisecond), I realized I was in pretty good company. One hundred eighty-three other countries had fallen for the bait and were obviously just as gullible as me. Thanks to the diligence and superior logic of congressmen like Sensenbrenner, our economy was not going to be crippled, and we could continue to pollute the planet in the manner to which we had become accustomed.

This perverse logic was strongly endorsed by the Bush administration, which not only refused to consider limits on carbon emissions but also argued that the United States was in fact *already* reducing its emissions, without any prompting from Kyoto. This bizarre idea flew in the face of easily observed facts, which showed that total U.S. emissions continued to climb, year by year, unabated. How could the president make such a claim?

Like a slick car salesman, Bush was being very selective in his use of words. From year to year the U.S. economy grows more or less continuously (notwithstanding the occasional economic collapse). This growth is measured in units of production, the "gross domestic product" or GDP. So GDP generally increases from one year to the next. But at the same time, we have devised methods of production that use less energy per unit of whatever we produce. Therefore, even though we have continued to burn more and more fossil fuel in total, and thus put more and more carbon dioxide into the atmosphere, we have reduced the amount of emissions *if we measure it in terms of each unit of GDP.* It's like having a barbecue every year and using more and more charcoal each time. You might grill six burgers one year, twelve the next, and twenty the year after. You could then claim that your use of charcoal had gone down, but only in terms of the amount used *per burger;* you still burned more charcoal each year. As far as the atmosphere is concerned, it's the total amount of fuel burned that is the crucial fact, not what you did with it.

The moral of the story is that when a used car salesman (or a president) says something that sounds too good to be true, you can be pretty sure that it is.

In fact, it's an even more devious argument than that. Over the last decade, we have "outsourced" much of our manufacturing to other countries, turning our economy more toward service activities—insurance, banking, computing, and so on. These produce economic wealth but do not require as much raw power as we used to consume when, for example, we produced a lot of steel and other industrial goods. Much of that activity now takes place in countries like China, and we import the goods we need from abroad. To power those industries, developing countries often rely on the least expensive fuel available— such as coal. It does not sit well with these countries when we happily import the cheap goods they manufacture for us and then turn around and demand that they clean up their act. These countries look to the developed world to take the lead in curbing carbon dioxide emissions while they try to raise the standard of living of their own citizens to the level we in the developed world have enjoyed for decades. They quickly point out that if you divide their total output by their population, their emissions per person are far lower than the amounts that developed countries produce. But again, the atmosphere has no interest in the way you justify the emissions. The carbon dioxide level just keeps going up, regardless.

The Kyoto Protocol did try to address these issues by setting up a mechanism in which the better-off countries provide funding to poorer countries to promote cleaner energy production and manufacturing facilities, more efficient transportation systems, and the like. The well-developed countries could also "offset" some of their emissions by paying for reforestation of other regions (particularly in the tropics), the idea being that if they could compensate for some unavoidable carbon dioxide emissions by reestablishing forests, that would effectively remove the same amount of carbon dioxide (or more) from the atmosphere. This is a sensible strategy with the added benefit of reversing the trend toward widespread deforestation of the tropics, with its attendant effects on biodiversity and soil erosion. It also recognizes the importance of forests in general, making clear-cutting and general degradation of forested regions an important political issue.

Critics claim that Kyoto was a failure, in part because the world's main producer of greenhouse gases, the United States, never signed on, but also because the reduction targets Kyoto set were too low and there were too many loopholes. These are valid criticisms, but in fact the Kyoto Protocol was a critically important agreement because for the first time, the greenhouse gas problem was universally acknowledged, and it set most nations of the world on the path toward a reduction of emissions. It is interesting to compare the response of the European Union with that of the United States. In the EU the public is now completely engaged in discussions about how to reduce their emissions of greenhouse gases. Television programs and magazine and newspaper articles explain how individuals can reduce their "carbon footprint" (how much fossil fuel they use). Businesses, schools, and government offices engage in friendly competitions to see who can reduce their energy use the most. Government policies have been directed toward investments and tax incentives to promote renewable sources of energy, such as solar, wind, and tidal power plants. Germany is now the world's major center for the manufacture of solar panels; half of the global production of electricity from solar panels occurs there. Denmark has increased its use of wind power to such an extent that some 20 percent of its energy production now comes from this source; its wind energy manufacturing industry is one of the country's biggest exporters, employing over 28,000 people.

Transportation has also become a critical part of the European debate. If you have to travel from point A to point B, information is available to help you decide whether it is better for the environment to go by car, train, or plane. Legislation has been introduced requiring that cars reduce the amount of carbon dioxide they produce. Consequently, all car advertisements now specify how many grams of carbon dioxide a particular model will emit per kilometer traveled. These are all remarkable changes that not only have led to a reduction of greenhouse gas emissions across Europe but also, more important, have set people on a path toward thinking about the energy they use and how that amount can be reduced.

In the United States such thinking has barely penetrated the general population, though there are some hopeful signs. In a few states such as California, Massachusetts, and Oregon, steps are well under way to promote green technologies, reduce energy use, and educate citizens. But at the same time, when California tried in December 2005 to mandate that vehicles sold in the state must meet certain (very modest) energy efficiency standards, the Environmental Protection Agency refused to provide the necessary waiver of federal Clean Air Act regulations and effectively prevented this from happening. The federal government has been impotent on the issue of energy efficiency. Energy efficiency standards for vehicles have remained static since Ronald Reagan was president, and to many in Congress the idea of reducing fuel use is downright "anti-American," equivalent to "crippling the economy." In fact, nothing could be further from the truth.

It is clearly better—for the economy and the environment—if we use less fuel in traveling from point A to point B. It is cheaper to make goods and provide services if we use less fuel. And it is good for the economy if we invest in research on renewable energy production, as this will lead to new industries, more jobs, and savings all around. Congressman Jay Inslee, a Washington Democrat, has been a passionate proponent of this view. He laid out a strategy for America's energy future in his book (with Bracken Hendricks), *Apollo's Fire: Igniting America's Clean Energy Future*. As he noted in a hearing of the House Committee on Energy and Commerce:

> Per unit of gross domestic product, we use half as much energy as we did in 1973. You think about that. Since 1973, our economy produces twice as much domestic product with the same amount of energy that it did in 1973, and there is just no reason on this green earth that all of a sudden we got stupid, that we are not going to be able to continue as the most brilliant society on Earth, and . . . continue those efficiency innovations. And they are not rocket science. Three of my neighbors drive cars that have already reduced their transportation-related CO_2 by 50%. The Chairman talked about the need to reduce our emissions by 40% to meet Kyoto. . . . it is well past the date where we need to move to solutions rather than debating the problem. . . . I am an optimist. . . . [B]ecause of our intellectual ability [we can] invent our way out of this pickle. Those who are people of great faith [are] becoming engaged in this debate because we are stewards of God's creation, and they are starting to urge Congress to act as well. We ought to be optimists and believe we can do it.[5]

This thinking is slowly sinking in. Today most of the developed countries of the world, as well as the less developed ones, are determined to do everything they can to develop renewable resources, reduce energy inefficiencies, and promote energy conservation. Kyoto, for all its limitations, was a milestone toward making that happen. There is still a long way to go, and contentious discussions over further reducing greenhouse gas emissions continue, but the direction has been set and there is no going back. We can all be thankful to the Kyoto negotiators for that.

As Al Gore nicely put it in his Nobel Peace Prize acceptance speech:

> The future is knocking at our door right now. Make no mistake, the next generation will ask us one of two questions. Either they will ask,
> *"What were you thinking; why didn't you act?"*

Or they will ask instead,
"How did you find the moral courage to rise and successfully resolve a crisis that so many said was impossible to solve?"[6]

The Kyoto Protocol gives us reason to believe that some nations have indeed grasped the magnitude of the problem and have taken steps to try to avoid a potential crisis. It may not have been a perfect plan, but it did provide a very important starting point for action.

The debate in the United States has moved further along, as people begin to understand that climate change is just part and parcel of much larger issues involving economic development and the creation of new "green" industries, national security issues, and environmental protection. The fact that many of the biggest new companies specializing in renewable energy technologies are based in Europe and Asia rather than North America has come as a surprise to many in Congress who expect the United States to be the technological leader in most fields. While this is certainly true in biomedical research, computing, and telecommunications, the United States has slipped far behind in developing the clean energy technologies of the future. The problem of protecting our oil supplies in volatile areas such as the Middle East, where we have to deal with corrupt governments and dictators to ensure that our oil supplies are secure, has also presented an unsettling picture to American voters. As the price of oil increases and our economy falters, we are sending more and more money abroad, often to pretty unsavory characters. In effect, we are borrowing money from foreign countries (mainly China) to import oil from other foreign countries so we can burn it (wastefully) in the United States. Add to this picture environmental disasters such as the BP oil spill in the Gulf of Mexico, which has created problems that will affect ecosystems and communities for many years to come. It is clear that the relentless pursuit of oil to feed our insatiable demand for fossil fuel (captured in the Republican campaign slogan "Drill, baby, drill") is a strategy that makes no sense and requires urgent action to change course.

There are some hopeful signs. As I write (June 2010), legislation to control greenhouse gases (by putting a price on carbon emissions) and to promote green energy technologies is being debated in the Senate, after passing the House of Representatives. President Obama has argued that this is an essential plank in his program to promote economic growth. Few politicians are willing to make the argument for limiting greenhouse gases on the basis of concern over climate change, given the more urgent concerns that the public has about jobs and economic uncertainties. But as long as the ship of state is turned in the right direction, I am not too worried about motivations or political arguments. As the former Chinese leader Deng Xiaoping famously said, "it doesn't matter if a cat is black or white, so long as it catches mice."

7
The Doubt Merchants
Suppression of Science and Character Assassination

Scientists, on the whole, are not alarmists. Scientific research is not aimed at finding results that set off warning bells. Scientists, in my experience, are a hardworking group of people who measure, observe, analyze, and report their findings in as clear a manner as they can. Careers depend on being objective, so that one's own results can be compared without exaggeration or hyperbole with what others have found. Extreme views are generally treated with skepticism until further investigations can confirm those concerns. And if confirmation is not found, it leaves the alarmists with egg on their faces. Reputations are simply not made by issuing shrill cries and inflated claims. Nevertheless, scientists do have a responsibility to report on the implications of their findings, especially if they seem to be of broad concern. Placing their results in a larger framework is appropriate, as long as the limitations and uncertainties of the research are made clear.

Problems arise when a pressing scientific issue runs up against major financial interests. And there are many scientific issues, especially those that relate to the environment, that have financial implications—on the upside for some and on the downside for others. If they point to the need for government action, the lines are quickly drawn on both sides of an issue. This is fertile ground for lobbyists and special interests to ensure that their clients are well represented.

Global warming is a perfect example of such an issue. To the scientists involved in climate research, or to those who study ecosystems or glaciers or the oceans, or any other aspect of the natural environment that may be affected by human-induced climate changes, the story is pretty clear. Mankind has inadvertently embarked on a gigantic global experiment, modifying the composition of the atmosphere at a faster rate than anything we can find evidence for in the past, leading to greenhouse gas concentrations that have not been seen on earth for millions of years. As far as we can judge, these changes will lead to a significant increase in global temperature and to an alteration of rainfall and snowfall patterns all over the world. The combination of rapid climate change, a burgeoning world population, and the associated destruction of the world's natural ecosystems is a potent mix. Some would call it a disaster; to others it's a catastrophe.

The passions of those convinced of the problem run high. They demand action and turn to their elected officials to address the matter without delay. But there are also those who are suspicious of science, who seem to be predisposed to disbelief. Their suspicions are well understood by those who represent the entrenched financial interests of the people and corporations that might be affected by government action. Those who oppose such action are skilled at putting the issues in simple but often misleading terms, and in recent years the efforts of such groups to stifle views that differ from their own have become more insidious. Not content with providing counterarguments (however misleading), they now commonly target the individuals who convey the message from the other side of the fence. There are no limits in these battles if the financial stakes are high enough. Character assassination has become a weapon in the arsenal of special interests. And if the stakes are really high, the pressure to alter or suppress information that may persuade the public to oppose the financial interests of the few reaches right into the government itself.

This was the situation during the presidency of George W. Bush. With a right-wing administration in place and Republicans controlling both houses of Congress, the stage was set for unrestrained exploitation of power. Instead of holding "hearings" (as in, let's hear about the issues on a subject and maybe learn something), the new standard was more akin to "tellings" (as in, we're going to tell you what we want you to hear), followed by "spinnings" in order to promote a particular view. As time went on, the arrogance of power was unbridled. When Senator

Inhofe called upon Michael Crichton to testify at a hearing on global warming, all pretensions to scientific objectivity and credibility went out the window. But those in power could not have cared less.

While Congress was pretending to deal fairly with climate change, at the other end of Pennsylvania Avenue the Bush administration was systematically implementing measures to restrict communications from federal science agencies. When the Environmental Protection Agency (EPA) produced a report in early June 2004 that raised concerns about global warming, Bush's reaction was curt and dismissive: "I read the report put out by the bureaucracy." Many scientists found that their reports about global warming were being edited to minimize any statements that might raise too much public concern. Interactions with the media were curtailed. Jim Hansen, head of NASA's Goddard Institute of Space Studies in New York, was a particular pain in the side of the Bush administration. He was viewed as a friend of Al Gore's and a potential problem because of his prominence and credibility as a scientist, and because he was not somebody who was easily intimidated. Hansen stood his ground, and when necessary he would take leave from his government position in order to speak as a private citizen, expressing his own views so as to free himself from government restrictions on what he could say.

At the same time all of this was going on, Vice President Dick Cheney was meeting with energy company executives to shape an energy policy for the country—without any input from those who might be considered "environmentalists." Efforts to find out who attended that meeting were resisted by the administration all the way to the Supreme Court, under the guise of "executive privilege." We never did find out who was part of that inner circle of advisers. Why the secrecy? What was it that we, the public, should not be told?

A cloud of suppression, obfuscation, and outright intimidation descended on the scientific establishment during the years of the Bush administration. My own experience of this was eye-opening. I had been asked to participate in a study of climate extremes: Have extreme weather conditions changed in frequency and magnitude over time? If so, what might that be due to? Could we expect extreme conditions to become more or less frequent in the future? That seemed an interesting set of questions, and I was happy to participate. My only reservation was that this effort was part of the government's Climate Change Science Program (CCSP). To me this was nothing more than a make-work program for federal scientists, who received instructions from on high to give the climate change issue some further study. Not satisfied with the Intergovernmental Panel on Climate Change, which obviously came to conclusions it did not want to hear, the Bush administration had commissioned the National Academy of Sciences to conduct its own review—a sort of mini-IPCC. When this report came out and endorsed the findings of the IPCC, the president's minders decided that there were surely important questions that required additional study before legislation to control greenhouse gases could be considered. So they dreamed up the CCSP to keep the science establishment busy and to postpone until far into the future the day when scientists would once again raise the alarm on global warming. And it worked. The last CCSP report came out just after Bush and his team left the White House. They had managed to kick the can down the road, doing absolutely nothing to address the global warming problem for eight years.

So it was with mixed feelings that I agreed to participate in preparing the report on climate extremes. On the one hand, I saw it as the offshoot of a cynical political machine; but on the other hand, the scientists involved were all excellent, and I felt it would be interesting to work with them on the issues. I could simply ignore the political overtones and work on the project for the sake of its intrinsic scientific interest. And so I went along to the first meeting, in Aspen, Colorado, in July 2005, and joined in the discussions of what should be done, who would do what, and how long it would take.

A month or so later one of the project leaders, Chris Miller, a program manager for the federal government's National Oceanic and Atmospheric Administration (NOAA), happened to be in Amherst. We were having dinner when I brought up a funny rumor I had heard. "Hey, Chris, somebody told me that we'll all have to be fingerprinted to work on that CCSP report." I tossed out this remark like the joke I thought it was. Chris smiled and hesitated in his response. "But you are a Brit, right? You have dual citizenship, so we'll just count you as a foreigner." It took a second or two for this to sink in. "You mean it's true? That we all have to be fingerprinted? Wait a minute—only Americans have to be fingerprinted? All the noncitizens on the panel do not have to be fingerprinted?" Surely he must be joking. But it soon became clear that he was not. Chris was clearly uncomfortable in telling me this. As he may have expected, I was outraged.

"You mean that, in order for me to voluntarily give my time to this effort, you expect me to be fingerprinted?" I just couldn't believe it. "No, Chris. I do not want you to count me as a Brit. In this case I want you to consider me a U.S. citizen, and I want you to write me a letter requesting that I go and get fingerprinted. And you know what?" I asked. "I will refuse, and I'll send the letter to *Nature,* and *Science,* and the *New York Times* and *Boston Globe* and any other newspaper that will take it. I'll show people what a sorry state we have come to when scientists have to be fingerprinted in order to write a simple report on climate extremes. Jeez—this is not some top-secret plan for nuclear war. It's just climate change, for goodness sake." I wonder if the panel of energy industry executives convened by Dick Cheney was also asked for their fingerprints.

Chris was in an awkward position, but it was not him I was mad at. He was trying to do the best he could under the constraints imposed on him by the Bush administration. He knew that past reports had been considered simple advisory documents, which the government could edit and change as it saw fit. But making all the members of the committee temporary employees of the government (and hence requiring that we all be fingerprinted) ensured that the report would be produced as a government document, supposedly without having sections redacted or edited out.

In fact, the administration was gradually becoming adept at editing any government report it considered inimical to its own goals—whether it came from NOAA or NASA or the EPA. In a truly remarkable gesture of arrogance, in early December 2007 White House staffers refused to open an e-mail message from the EPA summarizing the agency's finding that global warming poses a danger to the public.[1] Presumably, if they never read it, they could claim they did not know anything about it. They even removed six pages of testimony prepared by Julie Gerberding, director of the Centers for Disease Control, when she was called upon to tell Congress what concerns the CDC might have about the health threats associated with global warming. They had no qualms about ignoring warnings or altering scientific findings. No science report by any government agency was immune from such oversight.

Sadly, I never heard from the Climate Extremes Panel again. Evidently it was decided that my participation would trigger too many problems. Better to keep a low profile and go with the flow. I can understand that decision but regret that more scientists did not stand up and refuse to play the game, to scream from the rooftops, "I'm mad as hell and I'm not going to take it anymore!" But they didn't, and although a very good report came out, another low point was reached in the relationship between science and politics.

Situations such as the one I experienced with Chris prevailed throughout the Bush years, and though some of the actors have since changed, the underlying issues have not altered very much. The financial stakes are as high as they ever were, and the major energy players will not simply fade away. The U.S. Congress has not yet passed legislation that will significantly reduce carbon emissions to the atmosphere, and so for those who do not accept the need for such action, the game is still very much in play.

With infinitely deep pockets, the energy industry supports lobbyists, contributes huge amounts to the reelection campaigns of its favored politicians, and funds right-wing special interest groups, which hire pseudoscientists to dissect, trivialize, and often simply misrepresent scientific research. It is not difficult to pull apart much of the science that is published. Scientists are generally cautious and unwilling to say that something is absolutely certain. They couch their findings in caveats. They may see a pattern emerging, but it is rarely crystal clear. They make an observation, which may prompt others to examine the issue in a new way, with new data, slowly refining the argument.

Building a scientific premise is a bit like building a wall. The bricks might be a bit shaky to start with. Some may fall down by themselves or be crushed by others, while still others may find themselves well supported and locked into the fabric of the structure, eventually becoming impregnable to further pressure. In this way science builds toward a new understanding, a few bricks at a time, surrounded by a lot of rubble. But the wall goes up, and the important thing is not to focus too much on the latest brick added. It may not survive. But if you stand back and look at the bigger picture, it is often very impressive and formidable. What the denigrators generally choose to do is attack the most recent brick and, increasingly, the bricklayer who placed it there. They mock the assumptions, ridicule the methods, scoff at the conclusions.

Climate models provide easy pickings for these scavengers. All models are a simplification of reality, yet they do provide remarkably good simulations of how the climate system works. But they never get everything right; so if you ignore all the good material and focus on the bad, it seemingly gives you a license to dismiss the entire effort. In fact, the very word "model" is used with contempt, as though a model were somehow just a game, devoid of meaning. Yet models underlie many aspects of our lives. Cars, planes, buildings, electronic devices are all designed with computer models before anything is built. They allow us to simulate how complex things work, to test ideas and change conditions without having to go back and rebuild an entire structure. Models are therefore essential tools in science and engineering.

Climate science uses models to understand how the climate system works, to test ideas about what caused climate to change in the past, and to assess how it may change in the future. The models show clearly how explosive eruptions affected global climate in the past, and we can verify their simulations with evidence from natural archives, things that were affected by, and left a record of, past climatic conditions—tree rings, ice cores, lake sediments, corals, and so on. Models can help us assess how changes in the position of the earth in relation to the sun (so-called orbital variations) affect global climate, and we can see clearly how those changes brought about the waxing and waning of the ice ages. Most important, we can see how the buildup of greenhouse gases has altered global climate up to the present, and we can project forward in time to estimate how things may change in the future (see figures 6 and 7).

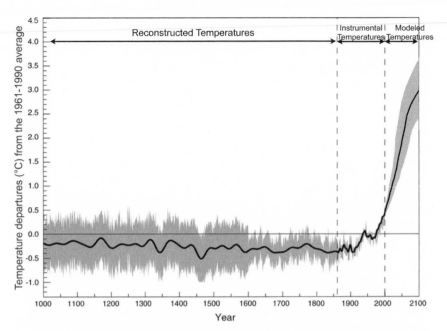

FIGURE 6. Projected change in global temperature (2000–2100) compared to instrumentally recorded temperatures from 1860 to 2000, and the reconstructed temperatures of the northern hemisphere for the period A.D. 1000 to A.D. 1860 (from Mann et al., 1999). The reconstructed temperatures and the recorded temperatures are the same as the figures 1 and 2 (pages 40 and 96). The projected temperatures are from the future scenario known as "A1B" in the IPCC report, which envisions carbon dioxide emissions rising to 16 billion metric tons by 2050 (compared to 2007 emissions of 8.5 billion) then declining to 13 billion by 2100. At that point in time atmospheric CO_2 levels would be around 650 ppm. This is a "middle of the road" estimate compared to the range of scenarios considered by the IPCC. The same scenario was used to assess changes in precipitation, in figure 7.

Source: Adapted from David S. Chapman and Michael G. Davis, "Climate Change: Past, Present, and Future," EOS 91 (2010): 325–26. (American Geophysical Union)

Having built up confidence in the ability of models to simulate today's climate as well as conditions in the past, we thus know that they are valuable tools for helping us understand how things are likely to change in the decades ahead. If these models are correct, we have strong reasons to be concerned, because they paint a picture of changes for which we are totally unprepared. Even with modest increases in greenhouse gases above present levels, temperatures will rise far beyond our

Projected Patterns of Precipitation Changes

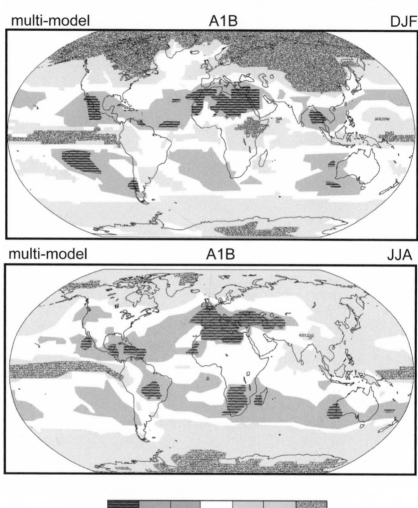

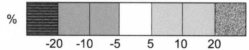

FIGURE 7. Expected changes in precipitation (rain and snow) in winter (top) and summer (bottom), based on the average of many computer model simulations, for a "mid-range" future climate scenario (A1B) compared to the average for 1980–1999.

Source: Adapted from Gerald A. Meehl et al., "Global Climate Projections," in Climate Change 2007: The Physical Science Basis. Contribution of Working Group I to the Fourth Assessment Report of the Intergovernmental Panel on Climate Change, *ed. Susan Solomon et al. (Cambridge: Cambridge University Press, 2007), 747–843.*

experience of the last thousand years. Furthermore, the additional energy in the climate system will bring about changes in the circulation of the atmosphere that will alter rainfall patterns, causing some areas to receive far less rainfall than they do today, while other areas will become much wetter. These simple statements about average conditions mask more important changes that we can expect, extreme events such as heat waves, severe thunderstorms, and strong hurricanes. No surprise, then, that insurance companies are at the front line of industries concerned about future climate changes, with the big reinsurance companies (SwissRe, MunichRe) actively promoting the importance of controlling greenhouse gases to limit the potential societal costs of more severe weather in the future.

Can we afford to ignore the concerns expressed by climate scientists, almost all of whom are convinced that we face a perilous future unless greenhouse gas emissions are significantly reduced? Or should we listen only to those representing the special interests, whose daily task is to pour scorn on the honest efforts of the scientific community? Should we not at least try to find a path that limits the worst case scenarios? Better yet, can we not devise solutions that provide us with multiple benefits—not just reduced emissions but new "green" industries, more jobs, economic growth, reestablishment of degraded ecosystems, reduced dependence on foreign energy supplies, and improvements in the standard of living for tens of millions of people around the world? Is such a vision compatible with our current standard of living?

I believe that we can make necessary changes, and that we must demand government action to address these issues. But in saying that, I just stepped from the domain of science into politics. I take this position on the basis of my understanding of the science of global warming. Science must remain separate from politics, but once scientists understand the issues, we must then decide our own political stance. By the same token, politics must stay out of science. Once politicians try to influence public opinion by manipulating scientific information or suppressing the findings of government scientists, we enter a world of duplicity and deception. Trust evaporates and cynicism triumphs. And then we all lose.

Today the hot topic may be global warming; tomorrow it could be evolution (again), or stem cell research, or studies of the human genome, or any number of a host of complex issues affecting the lives of millions and perhaps the wallets of a few. We need honest discussions, based on the best scientific information available, understanding that it may not be perfect and might even change as we learn more. But all political decisions have to be made on the basis of imperfect data; we rarely have all the facts, and this is obviously true as we try to anticipate how climate will change in the future. We also know, though, that all the decisions we make are based on a complex set of assumptions and estimations about how likely or unlikely some future occurrence might be. We don't know with any certainty that we will be in a car accident, or that our house will catch fire, but we have enough experience to know that there is a possibility that this might happen, and we take out insurance to limit the potential losses. We hedge our bets. And so it must be with future climate change.

The warning flags are up with regard to climate change. We should be paying attention and taking steps to prevent the dangers that lie ahead. Yes, the science may not be perfectly correct, and the predictions of looming disaster may turn out to be wrong. But the pendulum might swing even further in the other direction, toward a crisis that is worse than expected. Science is inherently conservative and generally cautious in its predictions. We should not place all our bets on the hope that future climate change scenarios are exaggerated. It is just as likely that they are a bit too rosy.

Whatever decisions we make, whatever strategies are chosen, they must be based on an honest and open debate, devoid of slurs against individuals and the distortion and misrepresentation of scientific information. Unfortunately, those opposed to any controls on greenhouse gas emissions have no interest in operating on such a level playing field. For them it is a "no holds barred" confrontation, and even criminal acts are fair play. This became apparent to me as I was watching CNN one evening in November 2009. There was a report that a server at the University of East Anglia's Climatic Research Unit (CRU) had been hacked into and a large number of private e-mail messages had been stolen. Suddenly a close-up appeared of a message sent to me, from Phil Jones, director of the CRU. I was shocked; that was private correspondence, not something for all the world to see! We are often warned that any electronic correspondence should be considered part of the public record, but nobody really expects their e-mail to be plastered all over the newspapers and broadcast on CNN. If we considered that likely, the frank and open exchange of ideas in modern societies (which is almost always performed via electronic media) would rapidly come to a screeching halt.

It turned out that the hacker had stolen thousands of e-mail messages from scientists at the Climatic Research Unit, dating back to 1996. This criminal act occurred just weeks before the Copenhagen UN climate summit, at which governments from all countries of the world were scheduled to try to hammer out an agreement on carbon dioxide emission reductions and a range of other issues related to global energy management. Needless to say, the stakes at such a meeting were extremely high, and the drumbeat of opposition from energy companies and their representatives was loud and persistent. They were fearful that an agreement might be reached that would effectively place a price on carbon as an atmospheric pollutant. Strangely, although this is a market-based approach to controlling carbon dioxide emissions, most energy companies reject the idea. Apparently they are all in favor of free-market capitalism unless it restricts how they operate, preferring instead to pass on the environmental costs of their emissions to the public at large.

But I digress. The stolen e-mail messages were quickly disseminated on the Internet, complete with handy software that enabled everyone to search quickly for "evidence" of fraud and deception. Within days the media, always looking for something controversial to fill the space between commercials, had dubbed the affair "Climategate." The word does have resonance in its allusion to Watergate, but there is little else to recommend it. In the episode known as Watergate, real crimes were committed in an attempt to influence elections, for which people went to prison; in this case, private correspondence was stolen and posted on the Internet with the sole purpose of embarrassing and discrediting the individuals concerned, and to divert attention from the important discussions that were about to take place in Copenhagen. The only similarity with Watergate was that it was an illegal action carried out for (im)purely political purposes, but the media hype that followed was focused not on the illegality of the theft but on the notion that the e-mail messages somehow revealed evidence of devious activity. A more careful reading of the e-mail shows this to be incorrect, though to anyone following the TV coverage, it was quite obvious that most reporters had never looked at the messages at all, but just relied on other poorly reported stories. In this way the media inflated and repeatedly misrepresented the facts. If there is a moral to be drawn from this sorry tale, it is that. After a couple of weeks of this bloodletting, the Associated Press finally assigned several reporters to read *all* the e-mail. They reported that there were about 1 million words in the stolen messages. Of these, a few sentences had been carefully selected by those looking for maximum press impact to create the impression of malfeasance. The other 999,000 or so words reveal nothing more than the daily grind of hardworking, dedicated scientists.

Regrettably, there were phrases in those e-mail communications that sound pretty bad in the cold light of a newspaper article. If you were sitting around drinking beer after a conference and you said, "That article is crap; I'm never going to cite it in the IPCC report!" you could expect to get a laugh. If you said, "I'm sick of people pestering me for data—I'll burn it before I give it up!" it would also sound pretty funny. Nobody would take it very seriously. But when you make the same remark in an e-mail message, which ends up plastered across the front page of *The Times* of London, it does look pretty awful, as Phil Jones is the first to admit. Obviously I can't reconstruct the context of all the messages that were leaked to the media, but it is worth noting that Phil was subjected to an extraordinary campaign of "Freedom of Information" (FOI) requests for data, orchestrated by Steve McIntyre, one of the original critics of our hockey stick paper. McIntyre has become a self-styled crusader for data, implying in his Internet blog that when data are not completely accessible to everybody, something shady is going on. Thus, when McIntyre discovered that some of the data used to compile the record of global average temperature, published by Phil Jones and others, were not available to him, he demanded that they be released. Phil's problem was that some of the data, provided by the meteorological agencies of a few countries, had come with the stipulation that they not be passed on except for academic purposes. That is, the agencies wanted to retain their option of selling the data but were agreeable to allowing it to be used for legitimate research purposes. Even though Phil explained this, McIntyre insisted that the data be released to the public. Through his website, he encouraged others to bury Phil in FOI requests. Consequently Phil Jones had to respond to dozens of demands—a chore so overwhelming that it completely swamped his research activities.

It was clear to Phil that these requests were frivolous and came largely from people who had no legitimate research interest in the data but were merely bent on harassing him. Nevertheless, each request required a certain amount of his time, and that led to some angry e-mail to his colleagues, expressing his frustration in rather intemperate language. Would I have acted any differently? Who knows. But if you read the blogs and news reports, you might think Phil Jones had committed mass murder. Not content with calls for his dismissal, some urged him to commit suicide, while others made more direct threats. Similar vitriolic e-mail messages were received by many of those with whom Phil had corresponded over the years (a large cross-section of the climatological community, given the importance of his research). Having gone through the Barton inquisition, I was somewhat prepared for the hysterical media onslaught, but I was completely taken aback by the venomous e-mail and letters that I and others received. I have no way of proving that this was a coordinated campaign, but the fact that so many of my colleagues received similar malicious correspondence makes me think that the drumbeat of manufactured outrage on the Internet and in the right-wing media was complicit in this activity.

I began to worry that this level of vitriol might be upsetting to Phil, whom I think of as a quiet and quite sensitive person. I have known him for thirty years, and we are good friends. So I decided to give him a call. No point in sending an e-mail! But I was quite unprepared for the conversation; he was depressed and having difficulty sleeping and eating because of the pressure. He could not go in to the office, as he would be hounded by reporters and protesters; he was a man in hiding. When I put the phone down, I was almost in tears. Phil is a great guy, as honest as the day is long, and his research has always been careful, thoughtful, and significant. The notion that he would purposely manipulate data to deceive anybody is completely ludicrous. In fact, an inquiry set up by the UK government to address that very question quickly concluded that research at the Climatic Research Unit was entirely aboveboard.[2] Yet here was the *Daily Telegraph* (a right-wing UK newspaper) declaring that the e-mail revealed "the worst scientific scandal of our generation."[3] Really? Has the world stopped warming? Did the sea level suddenly stop rising when the e-mail uproar broke? Have glaciers around the world all started to advance? The real scandal is the utterly inane level of discourse that this matter generated in the media. Commentators outbid one another in their expressions of shock. One day the e-mail messages were simply bad, the next day they were alarming, and by the end of the week they had become outrageous. Within a couple of weeks they had morphed into the greatest scientific scandal of our age. Well, you can't beat that for hyperbole. Hats off to the *Telegraph* for pushing the envelope to its limit, exposing the utter absurdity of the situation.

Much attention focused on a few phrases in the stolen e-mail, the most notorious of which read, "I've just completed Mike [Mann]'s *Nature* trick of adding in the real temps to each series for the last 20 years [i.e., from 1981 on] and from 1961 for Keith [Briffa]'s to hide the decline." This was from Phil Jones to several people, including me. To the recipients this was an uneventful note mentioning that he had simply combined the recent instrumental temperature record for the period after 1981 because most temperature reconstructions from paleo-records at the time ended around 1980. It was reasonable to use the instrumental record for the subsequent twenty years as representative of the longer period because the paleo-records tracked instrumental temperature extremely well up to then. An exception is tree ring records from a few high-latitude and high-elevation sites; these seem to have stopped tracking temperature in recent decades, and so obviously are not very helpful in showing the nature of temperature change. It is not clear why this is happening. It may be that recent warming has led to drier conditions, causing the trees to become more susceptible to moisture stress; or increases in anthropogenic particulate pollution may have limited photosynthetic activity in some locations. Whatever the reason, members of the Climatic Research Unit, led by Phil's colleague Keith Briffa, have been leaders in publishing articles about this problem as they try to figure it out, so they can hardly be accused of covering it up.

Phil could have explained all of this explicitly in his e-mail, but to his correspondents the context was clear and the words "trick" and "hide the decline" had no sinister meaning whatsoever. Not so to the denialists. To those eager to find a reason to dismiss anthropogenic global warming, this message was a gift from God. To them, it was prima facie evidence of fraud. Phil Jones had cooked the books and hidden evidence of global cooling in order to trick people. Furthermore, because Phil had divulged this plot in an e-mail message to his colleagues, none of whom had exposed this egregious act by turning him in to the thought police, all the recipients were guilty by association. Within days, websites appeared selling T-shirts with those dastardly words emblazoned across the front. Mocking videos showed up on YouTube.com, showing Mike Mann as a manipulative cartoon character. Character assassination was in full swing. Before long the issue had exploded into every media outlet, and "Climategate" was launched. Remarkably, CNN hosted not one but two hour-long specials and dispatched correspondents to Norwich to ferret out Phil Jones and confront him. Of course, this was simply cost-effective journalism because those correspondents were on their way to the Copenhagen climate summit anyway, to report on the negotiations to limit carbon dioxide emissions, yet few commented on the fact that these events were inextricably linked. Very soon, representatives from governments opposed to regulation took full advantage of the situation and expressed their outrage. Mohammad Al-Sabban from Saudi Arabia (and what is his country's major export?) was quick off the mark, demonstrating that he had no clue what the e-mail messages actually said when he told the BBC, "It appears from the details of the scandal that there is no relationship whatsoever between human activities and climate change."[4] He clearly comes from the school that believes if you say something that sounds profound, no matter how meaningless, it will become the truth.

By now the denialist industry could smell blood in the water. The gleeful fulminations on blogs and talk shows knew no bounds. Mike Morano, who served as Senator Inhofe's communications director on the Senate Public Works Committee from 2006 to 2009, remarked in a radio interview on April 22, 2010: "It is so nice to have the light of day and stench of corruption coming from people like Michael Mann and Rajendra Pachauri [head of the IPCC's Fourth Assessment Report] and Phil Jones and the upper echelon of UN scientists. We should be rejoicing that their entire careers are getting pissed on at the moment and justifiably so."[5] He was later quoted as saying: "I seriously believe we should kick them while they're down. They deserve to be publicly flogged."[6]

Regrettably, Morano was not the only one seeking retribution. Many of those who had corresponded with Phil Jones or Keith Briffa at the Climatic Research Unit, and whose e-mail had been stolen, were the subject of similar taunts on blogs and radio talk shows or via threatening e-mail and letters. Cyberbullying was in full swing, inflamed by those who sought to conceal their own view of the world behind their supposed outrage over the leaked messages. Consider this tirade of November 24, 2009, from Rush Limbaugh, the loudest mouth of right-wing radio:

> There are two worlds. We live in two universes. One universe is a lie. One universe is an entire lie. Everything run, dominated, and controlled by the left here and around the world is a lie. The other universe is where we are, and that's where reality reigns supreme and we deal with it. And seldom do these two universes ever overlap. A great illustration is what's happening here with what is now incontrovertibly known as a hoax. We know that the lead place, this Climate Research Unit at East Anglia University—which is

the number one adviser and communicator with the IPCC, which is the UN's climate-control crowd—we know that data was made up to advance the notion that man is causing the climate to warm. . . . So it's a hoax. We know these people—and I've known it all along. I know who these people are. I know who communists are. I know who liberals are. I know how they have to get things done. They have to lie. . . . [W]hat they have done here is now make it reasonable to doubt everything some scientist says who gets government money from somewhere. And if you know what's good for you, if you know that they're leftists, you won't believe anything they say anytime, anywhere, about anything. Their ideas are so hideous, are so insidious, so anti–free market, that they have to dress their ideas up in a phony cloak of compassion: Saving the planet, saving the polar bears, saving the water, saving the earth, saving whatever it is.[7]

After listening to rants like this, I feel like taking a shower to wash away the anger and venomous spittle that explodes from every sentence.

Thankfully, Limbaugh's views reflected only a small (albeit vocal) section of society, but the mainstream media were not much better. Mindless repetition of completely erroneous statements permeated every aspect of media coverage—in the press, on radio, and on television. What happened to careful investigative journalism? Apparently it died, and nobody seemed to care.

Unfortunately, "Climategate" was only the start of a new disinformation campaign. For those seeking to limit government action on carbon emissions, the main targets, as they had been for many years, were the IPCC reports. These had been given even greater status by the Nobel Committee's recognizing the IPCC, together with Al Gore, "for their efforts to build up and disseminate greater knowledge about manmade climate change, and to lay the foundations for the measures that are needed to counteract such change."[8]

Destroying the credibility of the IPCC reports was always the main prize, and the Copenhagen meeting provided the ideal venue for those efforts to be publicized. With little progress being made on an agreement to limit carbon emissions, the media were eager to find something they could sink their teeth into. And so "Glaciergate" was born, soon to be followed by "Amazongate" and several other "-gates." "Glaciergate" referred to a real error in one of the IPCC reports, which stated that the total area covered by glaciers in the Himalayas would decline from the present 0.5 million square kilometers to 0.1 million square kilometers by 2035, with damaging consequences for the downstream flow of rivers such as the Ganges. There is no strong evidence that this will occur, and it is clear that the IPCC made a mistake in reporting it as if there were. A second alleged scandal, dubbed "Amazongate," concerned a statement in the same report that "up to 40% of the Amazon forests could react drastically to even a slight reduction in precipitation. . . . [I]t is more probable that forests will be replaced by ecosystems that have more resistance to multiple stresses, caused by temperature increase, droughts and fires, such as tropical savannas."[9] This statement, with its embedded caveats, is not a mistake, though the citation given to support it was not an authoritative, peer-reviewed article as it should have been. Nevertheless, there are plenty of studies that point to the vulnerability of equatorial forests to future climate change, and so the statement is fully justified, on the basis of current research. But such subtleties were of no interest to the mainstream media. The opportunity to link an e-mail scandal with seemingly gross errors in the IPCC report was too good to miss, and so there was an outburst of almost hysterical coverage. A small industry broke out in which those seeking to raise maximum doubt about the credibility of the IPCC reports threw all their energy into dredging through every sentence and citation, trying to find even the most trivial error to expose the devious manipulations of those who had written them. And the media soaked it up, happy to announce each new triviality as "the scandal" dragged on.

In fact, the real story should have been: "Three Thousand–Page Report Contains Only a Couple of Errors!! Amazing Accuracy! Hundreds of Scientists Write an Almost Perfect Report!" But that is not the way the press works, a fact that is well understood by those who seek to confuse the public and sow doubt about the reality of anthropogenic global warming. As a result, further bloodletting ensued as the media wallowed in mock outrage. All the usual suspects took the opportunity to weigh in. Senator Inhofe, though no longer chairman of the Senate Committee on Environment and Public Works, issued a minority staff report exposing the "fraudulent" activities revealed by the East Anglia e-mail messages. Quickly going to the heart of his concerns, he noted: "The CRU controversy and recent revelations about errors in the IPCC's most recent science assessment cast serious doubt on the validity of EPA's endangerment finding for greenhouse gases under the Clean Air Act. The IPCC serves as the primary basis for EPA's endangerment finding for greenhouse gases." Inhofe was referring to the U.S. Supreme Court's decision in 2009 that the Environmental Protection Agency had the authority to regulate greenhouse gas emissions without congressional oversight.[10] In his concern that the White House might go ahead and issue limits on carbon pollution, using its authority over the EPA, he sought to discredit the IPCC and those associated with it. Not content, however, with issuing a report that might never see the light of day, Inhofe packaged it with a statement that was certain to get maximum media attention. He called for the indictment of seventeen climate scientists (including me) for, among other things, "the unlawful use of federal funds and potential ethical misconduct, involving violations of the Freedom of Information Act, the Shelby Amendment, Office of Science and Technology Policy Directives, President Obama's Transparency and Open Government Policy, Federal False Statements Act, The False Claims Act (Criminal) and Obstruction of Justice through Interference with Congressional Proceedings."[11]

Of course this had the desired effect. The media were all over it. None of the reports questioned the logic of Inhofe's selection of these seventeen scientists. Since we had all corresponded with Phil Jones at some point during the previous fifteen years, we were all clearly involved in something fishy that was worthy of a Justice Department inquiry. Fortunately, saner heads do occasionally prevail in Washington. After the initial burst of attention, the issue died away. But no matter. The main goal had been achieved: more stories in the press, on radio, and on television about the "ongoing scandal" over global warming. Anyone not paying complete attention could be forgiven for believing that there really was a problem. It was hardly surprising, then, that the ensuing stories all focused on how public confidence in climate scientists and the reality of global warming had rapidly declined. I discussed this with Beth Daley, who covers environmental issues for the *Boston Globe,* as I pointed out that writing more articles on this matter only served to reinforce the very story itself. "But the loss of confidence now *is* the news," she responded. And so the cycle was repeated.

All of this comes right out of the playbook of the "Merchants of Doubt," so well described in the book of that name by Naomi Oreskes and Erik Conway.[12] Unable to address adequately the bedrock scientific issues on which concerns over global warming are based, the doubt mongers chip away at the edges, focusing on minor issues, amplifying well-stated uncertainties, and, most perniciously, attacking the motivations and honesty of the scientists involved. This ploy began years ago with the attacks on Ben Santer (which have still not abated) and continued through the attempts to intimidate Mike Mann, Malcolm Hughes, and me; Phil Jones and other University of East Anglia faculty; IPCC head Rajenda Pachauri; and now seventeen U.S. scientists who have been branded as potential criminals. To the instigators of this relentless campaign, the goal is not prosecution. The goal is to ensure that legislation to control carbon emissions never passes into law. Character assassination, destruction of careers and reputations are no obstacle to them in the pursuit of their main goal.

This strategy has been applied before, with considerable success. By sowing the seeds of doubt, opponents delayed for decades legislation to reduce tobacco consumption, even though tobacco industry executives privately acknowledged the connection between tobacco use and cancer. More insidious still is the fact that many of the scientists who argued the case of the tobacco industry are the same people who are now disputing the significance of human-induced global warming. These experts-for-hire are not simply scientific prostitutes funded by energy companies and right-wing foundations to perform on their behalf. They commonly have a strong free market, conservative philosophy and view any sort of regulation as antithetical to their beliefs. It is therefore only a short step for them to equate any environmental concern with unnecessary government intervention, incipient socialism, or worse, communism. Consequently, regardless of the specific scientific issue of the day, they have attracted very significant financial support from right-wing foundations such as the Marshall Institute and wealthy conservative donors to defend the free market system. Whether it is cigarettes, acid rain, or global warming is apparently irrelevant. To them the answer is simple: free markets can resolve the problem better than government intervention, and they are more than willing to use their professional credentials to attack the scientific evidence in defense of market fundamentalism.

Of course, the same people usually have no problem with tax breaks or government subsidies for oil companies, or special legislation to limit product liability in the case of a lawsuit. Viewed through that lens, the notion of a completely free market is illusory. Nevertheless, too often these individuals are invited to appear on radio and television in order to present a seemingly balanced view of the science, generally without the interviewer explaining that what they are really representing is a political philosophy. So they tout their scientific credentials even though these may be completely unrelated to climate science. Regrettably, the general public often has a hard time weighing the arguments of two scientists and so concludes that there must really be some uncertainty about global warming. Thus the doubt merchants prevail, and public support for legislative action to address global warming declines. The strategy is well tested, very cynical, and, from the point of view of the energy industry, very effective. Legislation is stalled, diluted, or overturned. The winners are those who continue to pollute the atmosphere with impunity, while the losers are the rest of us, who will have to live with the consequences of unrestrained global warming.

As I bring this book to a close, the disinformation campaign rages on. In South Dakota in 2010, the state legislature passed a bill which "urges that instruction in the public schools relating to global warming include the following":

1. That global warming is a scientific theory rather than a proven fact;

2. That there are a variety of climatological, meteorological, astrological, thermological, cosmological, and ecological dynamics that can effect [sic] world weather phenomena and that the significance and interrelativity of these factors is largely speculative; and

3. That the debate on global warming has subsumed political and philosophical viewpoints which have complicated and prejudiced the scientific investigation of global warming phenomena.[13]

So "astrological" and "thermological" factors "effect" world weather. Who knew? No doubt the good citizens of South Dakota will appreciate the efforts of their legislators to protect their kids from contamination by real science.

A particularly pernicious attempt in the long series of character assassinations came from Ken Cuccinelli, the attorney general of Virginia, who issued a civil investigative demand (effectively a subpoena) to the University of Virginia, where Mike Mann worked from 1999 to 2005. In it he sought all e-mail communications, research materials, data, and computer codes relating to grants Mike had received while he was on the faculty there, to investigate whether Mike had violated the 2002 Virginia Fraud against Taxpayers Act, a law that was designed to prosecute individuals who make false claims in order to obtain government funds. It does not matter at all that there has never been any legal charge or evidence presented that Mike committed fraud. The mere fact that he was the main author on the "hockey stick" series of papers is apparently prima facie evidence that he must have defrauded the state government of its funds. Here is yet another example of a right-wing politician misusing his authority in order to intimidate an individual who delivered an unwelcome scientific message. Cuccinelli's abuse of power in pursuit of his own political agenda (what the journal *Nature* called his "ideologically motivated inquisition") set off many alarms. The American Association for the Advancement of Science issued a strong statement condemning Cuccinelli's action, asserting: "Scientists should not be subjected to fraud investigations simply for providing scientific results that may be controversial or inconvenient, particularly on high-profile topics of interest to society. . . . [I]nvestigations such as that targeting Professor Mann could have a long-lasting and chilling effect on a broad spectrum of research fields that are critical to a range of national interests. . . . Attorney General Cuccinelli's apparently political action should be withdrawn."[14]

This sentiment was echoed in numerous other statements from scientific organizations and newspaper editorials, all of which no doubt served to inflate Cuccinelli's high profile in the right-wing communities from which he seeks support. It is reassuring that the University of Virginia decided to stand up to his bullying tactics. In a powerful statement it rejected his request, noting:

Academic freedom is essential to the mission of our Nation's institutions of higher learning and a core First Amendment concern. As Thomas Jefferson intended, the University of Virginia has a long and proud tradition of embracing the "illimitable freedom of the human mind" by fully endorsing and supporting faculty research and scholarly pursuits. Our Nation also has a long and proud tradition of limited government framed by enumerated powers, which Jefferson ardently believed was necessary for a civil society to endure. The Civil Investigative Demands issued to the University by the Office of the Attorney General of Virginia threaten these bedrock principles. . . . [T]heir sweeping scope is certain to send a chill through the Commonwealth's colleges and universities. The Fraud Against Taxpayers Act . . . does not authorize the Attorney General to engage in scientific debate or advance the Commonwealth's positions in unrelated litigation about federal environmental policy and regulation. . . . Unfettered debate and the expression of conflicting ideas without fear of reprisal are the cornerstones of academic freedom; they consequently are carefully guarded First Amendment concerns. Investigating the merits of a university researcher's methodology, results, and conclusions (on climate change or any topic) goes far beyond the Attorney General's limited statutory power.[15]

Indeed, "unfettered debate and the expression of conflicting ideas without fear of reprisal are the cornerstones of academic freedom."

Alas, my experience of the last few years has shown me how fragile that concept is when the political and financial stakes are high. We must resist all those who seek to suppress scientific results that do not fit their preconceived ideas, and we must defend those whose reputations are attacked simply for reporting on science that may not be politically acceptable to those who wield power. Global warming and its associated environmental effects are not going away anytime soon. But the problem was created by mankind, and it can be solved by mankind. This is the challenge for us all, and one that I am confident we can overcome, provided we do not succumb to political intimidation and the dictates of denialists.

NOTES

1. The Congressional Hearings

1. For this and all quotations from the hearing, see *Senate Hearing 106-1115, Hearing before the Committee on Commerce, Science, and Transportation,* United States Senate, 106th Congress, 2nd Sess., May 17, 2000 (Washington, D.C.: GPO, 2003).
2. For the full text of Senator Inhofe's speech on January 4, 2005, see http://inhofe.sen ate.gov/pressreleases/climateupdate.htm.
3. A video of Inhofe's appearance can be seen at http://images1.americanprogress.org/ il80web20037/ThinkProgress/2006/inhofe.320.240.mov.
4. See www.opensecrets.org.
5. For this and all quotations from the hearing, see http://epw.senate.gov/hearing_state ments.cfm?id=246814.
6. For details of the letter, see www.ucsusa.org/scientific_integrity/abuses_of_science/ scientists-sign-on-statement.html.

2. A Letter from Congress

1. For the full text of the letter, see http://republicans.energycommerce.house.gov/ 108/ Letters/062305_Bradley.pdf.
2. According to a BBC News report (July 18, 2005), Myron Ebell said, "We've always wanted to get the science [of global warming] on trial," and "we would like to figure out a way to get this into a court of law," adding "This could work."

3. The Hockey Stick Controversy

1. Michael Mann, Raymond S. Bradley, and Malcolm K. Hughes, "Global Scale Temperature Patterns and Climate Forcing over the Past Six Centuries," *Nature* 392 (1998): 779–88.
2. Michael Mann, Raymond S. Bradley, and Malcolm K. Hughes, "Northern Hemisphere Temperatures during the Past Millennium: Inferences, Uncertainties, and Limitations," *Geophysical Research Letters* 26 (1999): 759–62.
3. Stephen McIntyre and Ross McKitrick, "Corrections to the Mann et. al. (1998) Proxy Data Base and Northern Hemispheric Average Temperature Series," *Energy & Environment* 14 (2003): 751–71.

4. Michael Mann, Raymond S. Bradley, and Malcolm K. Hughes, "Global Scale Temperature Patterns and Climate Forcing over the Last Six Centuries," *Nature* 430 (2004): 105.

5. For the full text of my letter, as well as those of Michael Mann and Malcolm Hughes, see www.geo.umass.edu/climate/bartonletter.html.

6. For a transcript of Galloway's testimony, see www.timesonline.co.uk/tol/news /world/article523583.ece.

7. For the full letter see http://www.geo.umass.edu/climate/waxman.pdf.

8. For a copy of the complete letter, see http://www.geo.umass.edu/climate/boehlert .pdf.

9. Thomas Hayden, "Science: Fighting over a Hockey Stick," *U.S. News and World Report,* July 14, 2005, www.usnews.com/usnews/culture/articles/050714/14climate .htm.

10. Andrew C. Revkin, "Two G.O.P Lawmakers Spar over Climate Study," *New York Times,* July 18, 2005, www.nytimes.com/2005/07/18/politics/18mann-final.html.

11. Juliet Eilperin, "Testy Exchange Reveals Rift in GOP over Global Warming," *Boston Globe,*July18,2005,wwwbostoncom/news/nation/washington/articles/2005/07/18 /testy_exchange_reveals_rift_in_gop_over_global_warming/.

12. See www.geo.umass.edu/climate/denverpostedit.pdf.

13. David Ignatius, "A Bill to Chill Thinking," *Washington Post,* July 22, 2005, www .washingtonpost.com/wp-dyn/content/article/2005/07/21/AR2005072102186.html.

14. Editorial, *New York Times,* July 22, 2005. The Center for Responsive Politics mentioned in this editorial is a nonpartisan research group that tracks money in politics. The editorial also noted that Barton worked in the oil and gas industry before being elected to Congress in 1984. Over the past decade he had consistently ranked as one of the top five recipients of campaign contributions from that industry.

15. Editorial, *Washington Post,* July 23, 2005.

16. For the complete statement, see www.egu.eu/statements/position-statement-of-the-divisions-of-atmospheric-and-climate-sciences-7-july-2005.html.

17. For the complete *Nature* editorial, see www.nature.com/nature/journal/v436/n70 47 /full/436001a.html.

18. For the full transcript of the letter, see www.aaas.org/news/releases/2005/0714 letter .pdf.

19. For the full transcript of the letter, see www.geo.umass.edu/climate/scientists-letter.pdf.

20. For the complete report, see www.nap.edu/openbook.php?record_id=11676&page =1.

21. Ironically, Mike was unable to make it because of severe weather.

22. The comparison of my text with that of the Wegman et al. report was kindly provided by Canadian blogger "Deep Climate": See http://deepclimate.files.wordpress .com/2009/12/wegman-bradley-tree-rings.pdf.

23. For Mann's complete response to questions from the House Energy and Commerce Subcommittee on Oversight and Investigations, see www.meteo.psu.edu/~mann /house06/HouseFollowupQuestionsMann31Aug06.pdf.

24. For transcripts and webcast of the hearing, see http://republicans.energycommerce. house.gov/108/hearings/07272006Hearing2001/hearing.htm.

25. The IPCC defines the term "very likely" to mean a probability in excess of 90 percent. If you see a weather forecast with a 90 percent probability of heavy rain, and you forget your umbrella, expect to get very wet.

26. See www.ipcc.ch/pdf/assessment-report/ar4/syr/ar4_syr_spm.pdf.

4. The IPCC and the Nobel Prize

1. For the history of the IPCC, see www.ipcc.ch/organization/organization_history. shtml.

2. For a further explanation of the IPCC organization, see www.ipcc.ch/working _groups/working_groups.shtml.

3. J. T. Houghton, L. G. Meira Filho, B. A. Callander, N. Harris, A. Kattenberg, and K. Maskell, eds., *Climate Change 1995: The Science of Climate Change* (Cambridge: Cambridge University Press, 1996), 4–5, summarizing points from the chapter "Detection of Climate Change and Attribution of Causes," by Benjamin D. Santer, Tom M. L. Wigley, Tim P. Barnett, and Ebby Anyamba, in the same volume; see the full chapter (407–43) for more extensive discussion.

4. Ibid.

5. For the complete text of all these letters, and how they were edited by the *Wall Street Journal* prior to their publication, see: www.ucar.edu/communications/quarterly /summer96/insert.html.

6. For the full context of this statement, see www.ipcc.ch/ipccreports/tar/vol4/english /008.htm.

7. For the full context of this statement, see www.ipcc.ch/publications_and_data/ar4 /wg1/en/spmsspm-understanding-and.html.

8. See www.ipcc.ch/ipccreports/tar/vol4/english/008.htm.

9. See http://nobelprize.org/nobel_prizes/peace/laureates/2007.

10. For the complete report, see www.cna.org/reports/climate.

5. Global Warming

1. See http://lightbucket.wordpress.com/2008/04/09/pr-versus-science-the-luntz memo.

6. Climate Futures

1. For the full text of the Convention, see http://unfccc.int/essential_background/con vention/background/items/1349.php.

2. See www.commondreams.org/pressreleases/Nov98/111298c.htm.

3. See http://inhofe.senate.gov/pressreleases/globalwarming.htm.

4. See http://clinton4.nara.gov/CEQ/earthday/ch3.html.

5. A full transcript is available at http://deepclimate.files.wordpress.com/2010/04 /hockey-stick-hearings-2006-ec-committee.pdf.

6. For the full text of Gore's speech, see http://nobelprize.org/nobel_prizes/peace /laureates/ 2007/gore-lecture_en.html.

7. The Doubt Merchants

1. Reported in the *New York Times,* June 25, 2008.

2. "We saw no evidence of any deliberate scientific malpractice in any of the work of the Climatic Research Unit"; www.uea.ac.uk/mac/comm/media/press/CRU statements/SAP.

3. *Daily Telegraph,* November 28, 2009.

4. See http://news.bbc.co.uk/2/hi/8392611.stm.

5. See www.climatesciencewatch.org/2010/04/30/denialist-morano-on-scientists -"rejoicing-that-their-entire-careers-are-getting-pissed-on/.

6. Quoted in Douglas Fischer, "Cyber Bullying Rises as Climate Data Are Questioned," *The Daily Climate,* March 1, 2010.

7. See www.rushlimbaugh.com/home/daily/site_112409/content/01125108.guest.html.

8. See http://nobelprize.org/nobel_prizes/peace/laureates/2007/press.html.

9. See www.ipcc.ch/publications_and_data/ar4/wg2/en/ch13s13-4.html#13-4-1.

10. See www.epa.gov/climatechange/endangerment.html.

11. See www.climatesciencewatch.org/2010/02/24/sen-inhofe-inquisition-seeking-ways -to-criminalize-and-prosecute-17-leading-climate-scientists/.

12. Naomi Oreskes and Erik M. Conway, *Merchants of Doubt: How a Handful of Scientists Obscured the Truth on Issues from Tobacco Smoke to Global Warming* (New York: Bloomsbury Press, 2010).

13. See http://legis.state.sd.us/sessions/2010/Bills/HCR1009P.htm.

14. The full statement is available at www.aaas.org/news/releases/2010/media/0518 board_statement_cuccinelli.pdf.

15. See www.virginia.edu/uvatoday/pdf/052710_petition.pdf.

RECOMMENDED READING

Archer, David. *Global Warming: Understanding the Forecast.* Oxford: Blackwell, 2007.

———. *The Long Thaw: How Humans Are Changing the Next 100,000 Years of Earth's Climate.* Princeton: Princeton University Press, 2009.

Archer, David, and Ray Pierrehumbert (Eds.). *The Warming Papers: The Scientific Foundation for the Climate Change Forecast.* Chichester: Wiley-Blackwell, 2011.

Archer, David, and Stefan Rahmstorf. *The Climate Crisis: An Introductory Guide to Climate Change.* Cambridge: Cambridge University Press, 2010.

Gore, Albert. *Our Choice: A Plan to Solve the Climate Crisis.* New York: Melcher Media, 2009.

Hansen, James. *Storms of My Grandchildren: The Truth about the Coming Climate Catastrophe and Our Last Chance to Save Humanity.* New York: Bloomsbury U.S.A., 2009

Hoggan, James (with Richard Littlemore). *Climate Cover-Up: The Crusade to Deny Global Warming.* Vancouver: Greystone-Books, 2009.

Inslee, Jay, and Bracken Hendricks. *Apollo's Fire: Igniting America's Clean Energy Economy.* Washington, D.C.: Island Press, 2008.

Kolbert, Elizabeth. *Field Notes from a Catastrophe: Man, Nature, and Climate Change.* New York: Bloomsbury U.S.A., 2006.

Mann, Michael E., and Lee R. Kump. *Dire Predictions: Understanding Global Warming; The Illustrated Guide to the Findings of the Intergovernmental Panel on Climate Change.* New York: DK Publishing, 2008.

Michaels, David. *Doubt Is Their Product: How Industry's Assault on Science Threatens Your Health.* Oxford: Oxford University Press, 2008.

Mooney, Chris. *The Republican War on Science.* New York: Basic Books, 2005.

———. *Storm World: Hurricanes, Politics, and the Battle over Global Warming.* Orlando: Harcourt, 2007.

Oreskes, Naomi, and Erik M. Conway. *Merchants of Doubt: How a Handful of Scientists Obscured the Truth on Issues from Tobacco Smoke to Global Warming.* New York: Bloomsbury Press, 2010.

Schmidt, Gavin, and Joshua Wolfe. *Climate Change: Picturing the Science.* New York: W. W. Norton, 2009.

Schneider, Stephen H. *Science as a Contact Sport: Inside the Battle to Save Earth's Climate.* Washington, D.C.: National Geographic, 2009.

Weart, Spencer. *The Discovery of Global Warming.* Cambridge: Harvard University Press, 2008.

INDEX

122, 123; and Nobel Peace Prize, 89, 90, 129–30, 149

greenhouse gases, 80, 105–9, 115–21; attempts to limit, 115–16, 123–31, 140 (*see also* Kyoto Protocol); defined, 101; and fossil fuel consumption, 1, 2, 12, 13–14, 30, 32, 91–92, 109–12, 116–17, 124, 126; and global warming, 32, 37, 48, 77–78, 91–92, 100–101, 109; unprecedented levels of, in atmosphere, 2, 12, 13, 49, 102, 113, 115–17. *See also* carbon dioxide

Greenland, 32, 107; and air pollution, 110, 114; ice cores from, 106–7, 114

Gulf of Mexico, 130

Hansen, Jim, 134

Hendricks, Bracken, 129

"hockey stick" study, 3; efforts to discredit, 3–4, 16–17, 18–19, 42–47, 71–74, 79 (*see also* Barton-Whitfield letters); *Geophysical Research Letters* and, 39, 44; IPCC and, 26, 41–42, 49–50, 51, 54, 66, 69, 79, 87–89; media coverage of, 29, 37–38, 39, 44, 61–64, 71; methodology and findings of, 30, 34–37, 39–41, 49, 97; National Academy of Sciences and, 47, 69–71; *Nature* editors and, 25, 37, 44, 46–47

Houghton, Sir John, 86

Hughes, Malcolm, 33, 56–57; attacks on, 19, 74 (*see also* Barton-Whitfield letters); and "hockey stick" study, 3, 19, 23, 28, 33, 70; and 2001 IPCC report, 42

Hurricane Katrina, 121

ice core records, 13, 35, 106–9, 110, 112, 114

ice shelves, 118

Ignatius, David, 63

India, 124–25

Indonesia, 124, 125

Industrial Revolution, 1, 30, 109–11, 114, 117, 124

Inhofe, James, 16–18, 76, 79, 123, 148, 150–51; and energy industry, 18, 62; on global warming as "hoax," 16, 42, 44; hearings held by, 18–19, 41, 133–34;

and "hockey stick" study, 16–17, 88; on "sound science," 16, 62

Inslee, Jay, 76, 128–29

insurance companies, 141

interglacial periods, 109, 117

Intergovernmental Panels on Climate Change (IPCC), 2–3, 77, 80, 83–87, 159n25; attacks on, 3, 26, 41–42, 69, 85–86, 89, 148–51; and Bush administration, 135; Fourth Assessment report of (2007), 79, 87, 139, 148; and "hockey stick" study, 26, 41–42, 49–50, 51, 54, 66, 69, 79, 87–89; and Nobel Peace Prize, 89, 90, 149; media coverage of, 41, 149–50; procedures of, 2–3, 41–42, 80–85; Third Assessment Report of (2001), 41, 42, 49–50, 79, 82, 87–89

internal combustion engines, 111

IPCC. *See* Intergovernmental Panels on Climate Change

island nations, 118

isotopes, 35, 106–9

Jackson, Andrew, 5

Jones, Phil, 143, 144–48, 151, 152

Jouzel, Jean, 107

Keeling, Charles, 105

Kerry, John, 8, 9–10, 11

Kiribati, 118, 122

Kyoto Protocol, 29, 122–30; U.S. Congress and, 11, 122–23, 124–25

Lane, Neal, 9, 11

Langway, Chet, 106–7

La Niña, 36

Lautenberg, Frank, 19–20

Leshner, Alan, 65–66

Limbaugh, Rush, 89, 90, 148–49

Lombardi, John, 28–29

London, 13, 37–38, 91

Lorius, Claude, 107

Luntz Research Company, 91

"Lysenkoism," 64–65

Mahlman, Jerry, 9, 39

Maldives, 118

Mann, Michael, 70, 72, 74–75, 146; attacks on, 17, 19, 147, 152, 154–56; background of, 33–34, 72, 74; and Barton-Whitfield letters, 23, 27–28, 51, 52–55, 59; and "hockey stick" study, 34, 36–37, 45, 70
Marshall Institute (George C. Marshall Institute), 85, 153
McCain, John, 8–9, 15
McCarthy, Joseph, 56
McIntyre, Stephen, 26, 43–45, 53, 54–55, 144–45
McKitrick, Ross, 26, 43–45, 53, 54–55
media coverage, 84–86, 151–53; of "Climategate," 143–45, 146, 147–49; of "hockey stick" controversy, 39, 41, 62–64, 71; of IPCC reports, 41, 149–51
"medieval warming period," 17–18, 32
Meira Filho, Luiz Gylvan, 86
methane (CH$_4$), 13, 101, 102, 116, 123
Milankovic, Milutin, 109
Military Advisory Board, 89–90
Miller, Chris, 135–36
modeling. See climate models
Molina, Mario, 66
Morano, Mike, 148
Moynihan, Daniel Patrick, 77

National Academy of Sciences, 20, 135; and "hockey stick" controversy, 47, 66, 69–71
National Center for Atmospheric Research, 53, 72
National Oceanic and Atmospheric Administration, 95, 135
National Science Foundation, 19, 23, 53, 58, 75
Nature, 65, 71, 154; handling of "hockey stick" study by, 25, 37, 44, 45–46
Neal, Larry, 61
New England, 99, 103, 104, 114
New Orleans, 99, 121
New York Times, 61, 63–64, 71
Nile River, 118
nitrous oxide (N$_2$O), 101, 102, 116, 123
Nobel Peace Prize, 89, 90, 129–30, 149
North, Gerald, 70

oceans, 99, 103, 119. See also ocean temperatures; sea level
ocean temperatures, 35, 36, 93, 97, 98, 117–18, 123
Oeschger, Hans, 106, 107
Ohio, 114
oil, 111, 124, 130. See also energy industry
Oreskes, Naomo, 152
Ozone Action, 8

Pachauri, Rajendra, 23, 26, 148, 152
Pacific Ocean, 17, 36
paleoclimatology, 34, 35, 73, 146–47. See also "climate proxies"
permafrost, 97
petroleum, 111
Philadelphia Inquirer, 63
pine beetles, 98, 99
plagiarism, 73–74
polar bears, 121–22
population growth, 1, 13, 110, 111–12, 133
power plants, 2, 114
precipitation changes, 140. See also rainfall patterns
"proxies." See "climate proxies"

rainfall patterns, 36, 119–20, 141
Raynaud, Dominique, 107
RealClimate Web site (www.realclimate.org), 46
renewable energy, 127–30
Royal Dutch Shell, 79

Santer, Ben, 3, 79, 83–87, 90, 152
Saudi Arabia, 147–48
Scandinavia, 32, 99
sea ice, 97, 99, 122
sea level, 77, 97, 99, 118; and coastal regions, 97, 118, 121; denialists and, 99; future prospects for, 118, 121
Seitz, Frederick, 85, 86
Senate Committee on Commerce, Science, and Transportation, 8–15
Senate Committee on Environment and Public Works, 16, 18–21, 148, 150–51. See also Inhofe, James
Sensenbrenner, Jim, 124–25

RAYMOND S. BRADLEY is a University Distinguished Professor in the Department of Geosciences and director of the Climate System Research Center at the University of Massachusetts Amherst. He did his undergraduate work at Southampton University (U.K.) and his postgraduate studies (M.S., Ph.D.) at the Institute of Arctic and Alpine Research, University of Colorado Boulder. He also earned a D.Sc. from Southampton University in 2003. His research interests are in climatology and paleoclimatology, with a particular focus on how climate has changed since the last ice age. He has written or edited twelve books on climatic change and written more than 180 articles on the topic.

He is a fellow of the American Geophysical Union, the American Association for the Advancement of Science, and the Arctic Institute of North America. In July 2006 he received an honorary doctorate from Lancaster University (Lancashire, U.K.). He was awarded the Oeschger Medal of the European Geosciences Union in 2007 and elected to the Finnish Academy of Science and Letters in 2008.